Advanced Approaches Applied to Materials Development and Design Predictions

Advanced Approaches Applied to Materials Development and Design Predictions

Special Issue Editors

Abílio M. P. De Jesus
José A. F. O. Correia
Shun-Peng Zhu
Xiancheng Zhang
Dianyin Hu

MDPI • Basel • Beijing • Wuhan • Barcelona • Belgrade • Manchester • Tokyo • Cluj • Tianjin

Special Issue Editors

Abílio M. P. De Jesus
University of Porto
Portugal

Xiancheng Zhang
East China University of Science
and Technology
China

José A. F. O. Correia
University of Porto
Portugal

Dianyin Hu
Beihang University
China

Shun-Peng Zhu
University of Electronic Science
and Technology of China
China

Editorial Office
MDPI
St. Alban-Anlage 66
4052 Basel, Switzerland

This is a reprint of articles from the Special Issue published online in the open access journal *Materials* (ISSN 1996-1944) (available at: https://www.mdpi.com/journal/materials/special_issues/ICMFM19).

For citation purposes, cite each article independently as indicated on the article page online and as indicated below:

LastName, A.A.; LastName, B.B.; LastName, C.C. Article Title. *Journal Name* **Year**, *Article Number*, Page Range.

ISBN 978-3-03928-412-2 (Pbk)
ISBN 978-3-03928-413-9 (PDF)

© 2020 by the authors. Articles in this book are Open Access and distributed under the Creative Commons Attribution (CC BY) license, which allows users to download, copy and build upon published articles, as long as the author and publisher are properly credited, which ensures maximum dissemination and a wider impact of our publications.

The book as a whole is distributed by MDPI under the terms and conditions of the Creative Commons license CC BY-NC-ND.

Contents

About the Special Issue Editors . . . . **vii**

Preface to "Advanced Approaches Applied to Materials Development and Design Predictions" **xi**

José Correia, Abílio De Jesus, Shun-Peng Zhu, Xiancheng Zhang and Dianyin Hu
Advanced Simulation Tools Applied to Materials Development and Design Predictions
Reprinted from: *Materials* **2020**, *13*, 147, doi:10.3390/ma13010147 . . . . **1**

Chun-Yi Zhang, Zhe-Shan Yuan, Ze Wang, Cheng-Wei Fei and Cheng Lu
Probabilistic Fatigue/Creep Optimization of Turbine Bladed Disk with Fuzzy Multi-Extremum Response Surface Method
Reprinted from: *Materials* **2019**, *12*, 3367, doi:10.3390/ma12203367 . . . . **5**

Monika Duda, Joanna Pach and Grzegorz Lesiuk
Influence of Polyurea Composite Coating on Selected Mechanical Properties of AISI 304 Steel
Reprinted from: *Materials* **2019**, *12*, 3137, doi:10.3390/ma12193137 . . . . **19**

Alexander Koch, Philipp Wittke and Frank Walther
Computed Tomography-Based Characterization of the Fatigue Behavior and Damage Development of Extruded Profiles Made from Recycled AW6060 Aluminum Chips
Reprinted from: *Materials* **2019**, *12*, 2372, doi:10.3390/ma12152372 . . . . **33**

Xin Liu, Zheng Liu, Zhongwei Liang, Shun-Peng Zhu, José A. F. O. Correia and Abílio M. P. De Jesus
PSO-BP Neural Network-Based Strain Prediction of Wind Turbine Blades
Reprinted from: *Materials* **2019**, *12*, 1889, doi:10.3390/ma12121889 . . . . **51**

Chunyi Zhang, Jingshan Wei, Huizhe Jing, Chengwei Fei and Wenzhong Tang
Reliability-Based Low Fatigue Life Analysis of Turbine Blisk with Generalized Regression Extreme Neural Network Method
Reprinted from: *Materials* **2019**, *12*, 1545, doi:10.3390/ma12091545 . . . . **67**

Grzegorz Lesiuk
Application of a New, Energy-Based ΔS^* Crack Driving Force for Fatigue Crack Growth Rate Description
Reprinted from: *Materials* **2019**, *12*, 518, doi:10.3390/ma12030518 . . . . **83**

Jian Chen, Chao Li, Jian Zhang, Cong Li, Jianlin Chen and Yanjie Ren
First-Principles Study on the Adsorption and Dissociation of Impurities on Copper Current Collector in Electrolyte for Lithium-Ion Batteries
Reprinted from: *Materials* **2018**, *11*, 1256, doi:10.3390/ma11071256 . . . . **97**

Sihai Luo, Liucheng Zhou, Xuede Wang, Xin Cao, Xiangfan Nie and Weifeng He
Surface Nanocrystallization and Amorphization of Dual-Phase TC11 Titanium Alloys under Laser Induced Ultrahigh Strain-Rate Plastic Deformation
Reprinted from: *Materials* **2018**, *11*, 563, doi:10.3390/ma11040563 . . . . **109**

Shaoxiong Xie, Jiageng Xu, Yu Chen, Zhi Tan, Rui Nie, Qingyuan Wang and Jianguo Zhu
Indentation Behavior and Mechanical Properties of Tungsten/Chromium co-Doped Bismuth Titanate Ceramics Sintered at Different Temperatures
Reprinted from: *Materials* **2018**, *11*, 503, doi:10.3390/ma11040503 . . . . **121**

Grzegorz Lesiuk, Michał Smolnicki, Dariusz Rozumek, Halyna Krechovska, Oleksandra Student, José Correia, Rafał Mech and Abílio De Jesus
Study of the Fatigue Crack Growth in Long-Term Operated Mild Steel under Mixed-Mode (I + II, I + III) Loading Conditions
Reprinted from: *Materials* **10.3390/ma13010160**, *2020*, 160, doi:10.3390/ma13010160 . . . . **135**

About the Special Issue Editors

Abílio M. P. De Jesus, since 2014, Dr. Abílio Manuel Pinho de Jesus has been currently auxiliary professor at the Department of Mechanical Engineering of the Faculty of Engineering from the University of Porto (FEUP), following 18 years of teaching activity at the Department of Engineering of the University of Trás-os-Montes e Alto Douro (UTAD), Vila Real, Portugal. He is also a research member at the Institute of Science and Innovation in Mechanical and Industrial Engineering (INEGI) and is an integrated research in the Associated Laboratory for Energy, Transports, and Aeronautics (LAETA). He graduated in Mechanical Engineering from FEUP in 1996. He received a master's degree in Mechanical Engineering from FEUP in 1999. He obtained a Ph.D. degree in Mechanical Engineering at UTAD in 2004. He is the co-author of more than 200 papers in national and international scientific journals (h-index = 25) and more than 300 papers presented and/or published into proceedings of both national and international conferences. His research activity has been developed in the fields of fatigue and fracture of materials and structures. He has focused on the manufacturing processes, and particularly subtractive and additive processes in the last 5 years. Process simulation, material characterization, hybrid manufacturing (additive and subtractive), and their relation with fatigue behavior are being targeted, besides other research topics related to the integrity of structures of different applications. His role as an editor of the Structural Integrity Book series, nomination for the ESIS TC12, and current European Projects, FASTCOLD and IN2TRACK2, and national projects MAMTOOL, ADD.STRENGTH and FIBREBRIDGE highlight his research in the domains of fatigue and manufacturing.

José A. F. O. Correia, researcher of CONSTRUCT/FEUP of the University of Porto (Portugal). Since 2018, he is a Guest Teacher at the Engineering Structures Department of the Civil Engineering and Geosciences Faculty of the Delft University of Technology (Netherlands). He is an Invited Assistant Professor at the structural mechanics section in the Civil Engineering Department of the University of Coimbra (since 2016/09). He obtained his BSc (2007) and MSc (2009) in Civil Engineering from the University of Trás-os-Montes e Alto Douro (UTAD). He is a specialist in steel and composite (steel and concrete) construction by the University of Coimbra in 2010. He is a Ph.D. in Civil Engineering at the University of Porto in 2015. He is also co-author of more 100 papers in the most relevant scientific journals devoted to engineering materials and structures and 200 proceedings in international and national conferences, congresses, and workshops. He is a member of scientific and professional organizations, such as Ordem dos Engenheiros, Associação Portuguesa de Construção Metálica e Mista (CMM), Associação para a Conservação e Manutenção de Pontes (ASCP), and European Structural Integrity Society (ESIS). He is Co-Chair of TC12 of ESIS, the Editor-in-Chief of the Springer Book Series Structural Integrity, and Guest Editor of several international journals. His current research interests are a) behavior to fatigue and fracture of materials and structures (steel and aluminum, riveted and bolted connections, pressure vessels, old steel bridges, wind turbine towers, offshore structures); b) probabilistic fatigue modeling of metallic materials (including statistical evaluation, size-effect, cumulative damage); c) probabilistic design of glass structural elements; d) mechanical behavior of materials and wooden structures (connections and characterization of ancient structures); e) mechanical and chemical characterization of old mortars and masonry structures.

Shun-Peng Zhu, Professor in Mechanical Engineering from the University of Electronic Science and Technology of China. He was an international fellow at Politecnico di Milano, Italy during 2016–2018 and research associate at the University of Maryland, the United States in 2010. His research which has been published in scholarly journals and edited volumes, over 100 peer-reviewed book chapters, journals and proceedings papers, explores the following aspects: fatigue design, probabilistic physics of failure modeling, structural reliability analysis, multi-physics damage modeling and life prediction under uncertainty, probability-based life prediction/design for engineering components. He received the Award of Merit of European Structural Integrity Society (ESIS)-TC12 in 2019, Most Cited Chinese Researchers (Elsevier) in the field of Safety, Risk, Reliability and Quality in 2018, 2nd prize of the National Defense Science and Technology Progress Award of Ministry of Industry and Information Technology of China in 2014, Polimi International Fellowship in 2015, Hiwin Doctoral Dissertation Award in 2012, Best Paper Awards of several international conferences and Elsevier Outstanding Reviewer Status. He serves as a guest editor and editorial board member of several international journals and Springer book series, Organizing Committee Co-Chair of QR2MSE 2013, TPC Member of QR2MSE 2014-2019, ICMR 2015, ICMFM XIX 2018-2020 and IRAS 2019.

Xiancheng Zhang received his Ph.D. degree from Shanghai Jiao Tong University, China in 2007. Then he moved to the National Institute for Materials Science (NIMS) in Japan to act as a post-doctoral researcher for 1 year. He has contributed considerably to life design and prediction methods of high-temperature components and to the development of advanced surface manufacturing techniques. He has published more than 100 peer-reviewed papers, including more than 70 SCI-indexed papers in such journals including Acta Materialia, Journal of Applied Physics, Engineering Fracture Mechanics, Surface and Coatings Technology. Dr. Zhang received a number of distinguished awards including International Institute of Welding (IIW) Granjon Prize, Shanghai outstanding doctoral dissertation award, nomination of Chinese outstanding doctoral dissertation award, and Chinese petroleum chemical industry association technological award, the first-class of Shanghai natural science prize, the first-class of Beijing natural science prize, and the second-class of national nature science prize of China. He was the recipient of the New Century Excellent Talents Program Award (2011) from the Ministry of Education of China, the Outstanding Young Talents Award (2012), Shanghai Pujiang Talent (2012), the National Science Fund for Excellent Young Scholars of China (2013), Education Award for Young Teachers by FOK YING TUNG Education Foundation from the Ministry of Education of China (2014), Shanghai Young Sci-tech Talents (2014), Changjiang Young Scholars Programme of China (2015), National Science Fund for Distinguished Young Scholars of China (2017).

Dianyin Hu is a Professor in the School of Energy and Power Engineering from the Beihang University of China. She was a post-doctoral researcher at McGill University, Canada during 2015–2016. She focuses on fatigue life prediction, probabilistic-based life evaluation, composites damage, multi-scale modeling, and multidisciplinary design optimization (MDO) for aero-engine components. She received the Award of Merit of European Structural Integrity Society (ESIS)-TC12 in 2019, Talented Young People of Beijing University Award in 2013, Best Paper Awards of several international conferences and Elsevier Outstanding Reviewer Status. She serves as guest editor of several international journals and Organizing Committee Co-Chair of WCCM—APCOM 2016, ICMFM XIX 2018–2020 and IRAS 2019.

Preface to "Advanced Approaches Applied to Materials Development and Design Predictions"

This Special Issue explores the limits of the current generation of materials, which are continuously being reached according to the frontier of hostile environments, whether in the aerospace, nuclear, or petrochemistry industry, or in the design of gas turbines where the efficiency of energy production and transformation demands has increased temperatures and pressures.

This Special Issue has attracted submissions from China, Poland, Germany, and Portugal: 17 submissions have been received and 10 articles were published.

Zhang's group from the Harbin University of Science and Technology, Fudan University and Northwestern Polytechnical University (China) developed an investigation entitled "Probabilistic Fatigue/Creep Optimization of Turbine Bladed Disk with Fuzzy Multi-Extremum Response Surface Method", where the probabilistic fatigue/creep coupling optimization of turbine bladed disks was implemented—the rotor speed, temperature, and density as optimization parameters, and the creep stress, creep strain, fatigue damage, and creep damage as optimization objectives.

Duda, Pach, and Lesiuk presented the paper "Influence of Polyurea Composite Coating on Selected Mechanical Properties of AISI 304 Steel", in which the results of an experimental campaign on mechanical characterization of the AISI 304 steel with composite coatings, where the impact of the applied polyurea composite coating on selected mechanical properties, mainly, adhesion, impact resistance, static behavior, and fatigue lifetime of notched specimens were researched.

Kotch et al. from the TU Dortmund University wrote a scientific work entitled "Computed Tomography-Based Characterization of the Fatigue Behavior and Damage Development of Extruded Profiles Made from Recycled AW6060 Aluminum Chips", where an investigation related with the quasi-static and cyclic mechanisms to identify the possible parameters that can influence the mechanical properties of extruded chip-based profiles, is suggested. In this research, the authors analyzed all specimens by X-ray computed tomography (CT) before the tests in order to be able to detect possible influences of defects like pores and delamination on the mechanical properties.

Liu et al. presented a study entitled "PSO-BP Neural Network-Based Strain Prediction of Wind Turbine Blades". These algorithms have an important advantage in dealing with non-linear fitting and multiple input parameters. Thus, these authors have established a strain-predictive PSO-BPNN model for a full-scale static experiment of a certain wind turbine blade.

Another study entitled "Reliability-Based Low Fatigue Life Analysis of Turbine Blisk with Generalized Regression Extreme Neural Network Method" was introduced by Zhang et al., where the generalized regression extreme neural network (GRENN) method was proposed by integrating the basic thoughts of generalized regression neural network (GRNN) and the extreme response surface method (ERSM).

Normally, fatigue crack growth relations are presented by using a linear-elastic stress intensity factor range, ΔK. Lesiuk, from the Wroclaw University of Science and Technology, has proposed a new energy-based crack driving force for the description of the fatigue crack growth rates.

Chen et al. from the Changsha University of Science and Technology and Guangxi University (China) presented a scientific work entitled "First-Principles Study on the Adsorption and Dissociation of Impurities on Copper Current Collector in Electrolyte for Lithium-Ion Batteries", where the stable configurations of HF, H2O, and PF5 adsorbed on Cu(111) and the geometric parameters of the admolecules were confirmed after structure optimization.

Luo et al. published a paper entitled "Surface Nanocrystallization and Amorphization of Dual-Phase TC11 Titanium Alloys under Laser Induced Ultrahigh Strain-Rate Plastic Deformation". An innovative surface technology, laser shock peening (LSP), to the dual-phase TC11 titanium alloy to fabricate an amorphous and nanocrystalline surface layer at room temperature, for the ultrahigh strain-rate plastic deformation, are applied.

Xie et al. presented a work entitled "Indentation Behavior and Mechanical Properties of Tungsten/Chromium co-Doped Bismuth Titanate Ceramics Sintered at Different Temperatures", where the indentation behavior, as well as the mechanical properties of tungsten/chromium co-doped bismuth titanate ceramics sintered at different temperatures, are addressed. According to this scientific work, lower hardness and higher fracture toughness was verified for high sintering temperature.

Finally, a Polish–Portuguese team presented a paper entitled "Study of the Fatigue Crack Growth in Long-Term Operated Mild Steel Under Mixed-Mode (I+II, I+III) Loading Conditions". An experimental campaign for evaluating the mixed-mode fatigue propagation behavior supported by numerical simulation was undertaken. Additionally, SEM analysis of fracture surfaces of the specimens was conducted.

Abílio M. P. De Jesus, José A. F. O. Correia, Shun-Peng Zhu, Xiancheng Zhang, Dianyin Hu
Special Issue Editors

Editorial

Advanced Simulation Tools Applied to Materials Development and Design Predictions

José Correia [1,*], Abílio De Jesus [2], Shun-Peng Zhu [3], Xiancheng Zhang [4] and Dianyin Hu [5]

1. CONSTRUCT, Department of Civil Engineering, University of Porto, 4200-465 Porto, Portugal
2. INEGI, Department of Mechanical Engineering, University of Porto, 4200-465 Porto, Portugal; ajesus@fe.up.pt
3. Center for System Reliability and Safety, University of Electronic Science and Technology of China, Chengdu 611731, China; zspeng2007@uestc.edu.cn
4. Key Laboratory of Pressure Systems and Safety, Ministry of Education, East China University of Science and Technology, Shanghai 200237, China; xczhang@ecust.edu.cn
5. School of Energy and Power Engineering, Beihang University, Beijing 100083, China; hdy@buaa.edu.cn
* Correspondence: jacorreia@inegi.up.pt or jacorreia@fe.up.pt

Received: 28 December 2019; Accepted: 30 December 2019; Published: 30 December 2019

Abstract: This thematic issue on advanced simulation tools applied to materials development and design predictions gathers selected extended papers related to power generation systems, presented at the XIX International Colloquium on Mechanical Fatigue of Metals (ICMFM XIX) organized at University of Porto, Portugal, in 2018. Guest editors express special thanks to all contributors for the success of this special issue—authors, reviewers, and journal staff.

Keywords: damage/degradation; failure mechanisms; probabilistic physics; advanced testing and statistics; materials technology; power generation systems and technologies

1. Introduction

Fatigue damage represents one of the most important degradation phenomena which structural materials are subjected to in normal industrial operation, which may finally result in a sudden and unexpected failure/fracture. Since metal alloys are still the most used materials for the design of the majority of components and structures intended to carry out the highest service loads, the study of the different aspects of metals fatigue still attracts the permanent attention of scientists, engineers, and designers.

The first International Colloquium on Mechanical Fatigue of Metals (ICMFM) was organized in Brno, Czech Republic in 1968. Afterwards, regular Colloquia on Mechanical Fatigue of Metals started in 1972 also in Brno and were originally limited to participants from the countries of the former "Eastern Block". They continued until the 12th Colloquium in 1994 at Miskolc, Hungary, every two years. After a break twelve years long, the Colloquia restarted in 2006 at Ternopil, Ukraine, followed by the ones in 2008 (Varna, Bulgaria), 2010 (Opole, Poland), 2012 (Brno, Czech Republic), 2014 (Verbania, Italy) [1], and 2016 (Gijón, Spain) [2]. The last two organizations indented to open the Colloquium to participants from all countries across Europe interested in the subject of fatigue of metallic materials [3,4]. The XIX International Colloquium on Mechanical Fatigue of Metals (ICMFM XIX) [5] was organized in 5–7 September 2018, at the Faculty of Engineering of the University of Porto, in Porto City, located at seaside in the northwest region of Portugal. This International Colloquium was intended to facilitate and encourage the exchange of knowledge and experiences among the different communities involved in both basic and applied research in the field of fatigue of metals, exploring the problem with a multiscale perspective, using both analytical and numerical approaches, without losing the perspectives of the applications [5–8].

This special issue approaches the thematic about the limits of the current generation of materials, which are continuously being reached according to the frontier of hostile environments, whether in the aerospace, nuclear, or petrochemistry industry, or in the design of gas turbines where efficiency of energy production and transformation demands increased temperatures and pressures [6]. At the same time, increasing the performance and reliability, in particular by controlling and understanding of early failures, is one key point for future materials. Moreover, increasing material lifetimes in service and the extension of recycling time are expected. Accordingly, continued improvements on "materials by design" have been possible through accurate modeling of failure mechanisms by introducing advanced theoretical and simulation approaches/tools. Based on this, researches on failure mechanisms can provide assurance for new materials at the design stage and ensure the integrity in the construction at the fabrication phase. Specifically, material failure in hostile environments occurs under multiple sources of variability, resulting from environmental load, material properties, geometry variations within tolerances, and other uncontrolled variations. Thus, advanced methods and applications for theoretical, numerical, and experimental contributions that address these issues on failure mechanism modeling and simulation of materials are desired and expected.

2. Scientific Topics

This issue collects selected papers from ICMFM XIX [5] related to advanced analytical and numerical simulation approaches applied to materials development and design predictions, about power generation systems. The scientific topics addressed in this issue are summarized as follows:

- Environmental assisted fatigue;
- Multi-damage/degradation;
- Multi-scale modeling and simulation;
- Micromechanics of fracture;
- Material defects evolution;
- Interactions of extreme environments;
- Microstructure-based modeling and simulation;
- Fracture in extreme environments;
- Probabilistic physics of failure modeling and simulation;
- Probabilistic optimization;
- Advanced testing and simulation;
- Life prediction and extension;
- Stochastic degradation modeling and analysis;
- Low- and high-cycle fatigue;
- Artificial intelligence methods.

3. Overview on the Themed Issue

This section addresses in brief the scientific papers published in this thematic issue on advanced analytical and numerical simulation approaches applied to materials development and design predictions about power generation systems.

Zhang's group from the Harbin University of Science and Technology, Fudan University and Northwestern Polytechnical University (China) developed a fuzzy multi-extremum response surface method (FMERSM) for the comprehensive probabilistic optimization of multi-failure/multi-component structures, where the probabilistic fatigue/creep coupling optimization of turbine bladed disks was implemented—the rotor speed, temperature, and density as optimization parameters, and the creep stress, creep strain, fatigue damage, and creep damage as optimization objectives [9].

Duda, Pach, and Lesiuk [10] presented results of an experimental campaign on mechanical characterization of the AISI 304 steel with composite coatings, where the impact of the applied polyurea composite coating on selected mechanical properties, mainly, adhesion, impact resistance, static behavior, and fatigue lifetime of notched specimens was researched.

Kotch et al. [11] from the TU Dortmund University investigated the quasi-static and cyclic mechanisms to identify the possible parameters that can influence the mechanical properties of extruded chip-based profiles. In this research, the authors analyzed all specimens by X-ray computed tomography (CT) before the tests in order to be able to detect possible influences of defects like pores and delamination on the mechanical properties.

Liu et al. [12] suggested a study on a new strain prediction method by introducing intelligent algorithms—back propagation neural network (BPNN) improved by Particle Swarm Optimization (PSO). These algorithms have an important advantage in dealing with non-linear fitting and multiple input parameters. Thus, these authors have established a strain-predictive PSO-BPNN model for full-scale static experiment of a certain wind turbine blade.

Other study related to the influence of thermal–structural coupling on the blisk low-cycle fatigue life reliability analysis was introduced by Zhang et al. [13], where the generalized regression extreme neural network (GRENN) method was proposed by integrating the basic thoughts of generalized regression neural network (GRNN) and the extreme response surface method (ERSM).

Normally, fatigue crack growth relations are usually presented by using a linear-elastic stress intensity factor range, ΔK. Lesiuk [14], from the Wroclaw University of Science and Technology, has been proposed a new energy-based crack driving force for the description of the fatigue crack growth rates.

Chen et al. group from the Changsha University of Science and Technology and Guangxi University (China) presented a scientific work entitled by first-principles study on the adsorption and dissociation of Impurities on Copper Current Collector in Electrolyte for Lithium-Ion Batteries, where the stable configurations of HF, H_2O, and PF_5 adsorbed on Cu(111) and the geometric parameters of the admolecules were confirmed after structure optimization [15].

Luo et al. research team applied an innovative surface technology, laser shock peening (LSP), to the dual-phase TC11 titanium alloy to fabricate an amorphous and nanocrystalline surface layer at room temperature, for the ultrahigh strain-rate plastic deformation [16].

Xie et al. [17] have presented the indentation behavior, as well as the mechanical properties, of tungsten/chromium co-doped bismuth titanate ceramics sintered at different temperatures. According to this scientific work, a lower hardness and a higher fracture toughness was verified for high sintering.

Finally, a Polish–Portuguese team [18] presented an analysis of the mixed-mode (I+II, I+III) fatigue crack growth rates in bridge steel after 100-year operating time. An experimental campaign for evaluating the mixed-mode fatigue propagation behavior supported by numerical simulation was undertaken. Additionally, SEM analysis of fracture surfaces of the specimens was conducted.

4. Final Remarks

Guest Editors for this thematic issue are pleased with the final result of the published papers and hope that these scientific works can be useful to researchers, engineers, designers, and other colleagues involved in different thematic aspects of the advanced analytical and numerical simulation approaches applied to materials development and design predictions about power generation systems.

Additionally, the Guest Editors would like to express gratitude to all authors for their contributions and to all reviewers for their generous work that is fundamental in the dissemination of the scientific finding.

Finally, the Guest Editors would also like to express special thanks to the Editorial Board of Materials international journal for help, support, and patience and for their exceptional contributions during this time.

Acknowledgments: As the Guest Editors, we would like to thank all the authors who submitted papers to this Special Issue. All the papers published were peer-reviewed by experts in the field whose comments helped to improve the quality of the edition. We also would like to thank the Editorial Board of *Materials* for their assistance in managing this Special Issue.

Conflicts of Interest: The authors declare no conflict of interest.

References

1. Guagliano, M.; Vergani, L. Foreword 17th International Colloquium on Mechanical Fatigue of Metals, ICMFM 2014. *Procedia Eng.* **2014**, *74*, 1. [CrossRef]
2. Pariente, I.F.; Belzunce, J.; Canteli, A.F.; Correia, J.A.F.O.; De Jesus, A.M.P. Editorial of the ICMFM Conference. *Procedia Eng.* **2019**, *160*, 1–4. [CrossRef]
3. Correia, J.A.F.O.; De Jesus, A.M.P.; Pariente, I.F.; Belzunce, J.; Fernández-Canteli, A. Mechanical fatigue of metals. *Eng. Fract. Mech.* **2017**, *185*, 1. [CrossRef]
4. De Oliveira Correia, J.A.F.; Calvente, M.M.; De Jesus, A.M.P.; Fernández-Canteli, A. Guest editorial. *Int. J. Struct. Integr.* **2017**, *8*, 614–616.
5. Correia, J.A.O.C.; De Jesus, A.M.P.; Fernandes, A.A.; Calçada, R.A.B. Mechanical Fatigue of Metals-Experimental and Simulation Perspectives. In *Structural Integrity*; Correia, J.A., De Jesus, A.M.P., Eds.; Springer: Berlin, Germany, 2019; Volume 7, p. 413. ISBN 978-3-03-013979-7.
6. Correia, J.A.F.O.; De Jesus, A.M.P.; Muniz-Calvente, M.; Sedmak, A.; Moskvichev, V.; Calçada, R. The renewed TC12/ESIS technical committee-Risk analysis and safety of large structures and components. *Eng. Fail. Anal.* **2019**, *105*, 798–802. [CrossRef]
7. Correia, J.A.F.O.; Berto, F.; Ayatollahi, M.; Marsavina, L.; Kotousov, A.; Sedmak, A. Guest editorial: Advanced design and fatigue assessment of structural components. *Fatigue Fract. Eng. Mater. Struct.* **2019**, *42*, 1217–1218. [CrossRef]
8. Correia, J.A.F.O.; Lesiuk, G.; De Jesus, A.M.P.; Calvente, M. Recent developments on experimental techniques, fracture mechanics and fatigue approaches. *J. Strain Anal. Eng. Des.* **2018**, *53*, 545. [CrossRef]
9. Zhang, C.-Y.; Yuan, Z.-S.; Wang, Z.; Fei, C.-W.; Lu, C. Probabilistic Fatigue/Creep Optimization of Turbine Bladed Disk with Fuzzy Multi-Extremum Response Surface Method. *Materials* **2019**, *12*, 3367. [CrossRef] [PubMed]
10. Duda, M.; Pach, J.; Lesiuk, G. Influence of Polyurea Composite Coating on Selected Mechanical Properties of AISI 304 Steel. *Materials* **2019**, *12*, 3137. [CrossRef] [PubMed]
11. Koch, A.; Wittke, P.; Walther, F. Computed Tomography-Based Characterization of the Fatigue Behavior and Damage Development of Extruded Profiles Made from Recycled AW6060 Aluminum Chips. *Materials* **2019**, *12*, 2372. [CrossRef] [PubMed]
12. Liu, X.; Liu, Z.; Liang, Z.; Zhu, S.-P.; Correia, J.A.F.O.; De Jesus, A.M.P. PSO-BP Neural Network-Based Strain Prediction of Wind Turbine Blades. *Materials* **2019**, *12*, 1889. [CrossRef] [PubMed]
13. Zhang, C.; Wei, J.; Jing, H.; Fei, C.; Tang, W. Reliability-Based Low Fatigue Life Analysis of Turbine Blisk with Generalized Regression Extreme Neural Network Method. *Materials* **2019**, *12*, 1545. [CrossRef] [PubMed]
14. Lesiuk, G. Application of a New, Energy-Based ΔS* Crack Driving Force for Fatigue Crack Growth Rate Description. *Materials* **2019**, *12*, 518. [CrossRef] [PubMed]
15. Chen, J.; Li, C.; Zhang, J.; Li, C.; Chen, J.; Ren, Y. First-Principles Study on the Adsorption and Dissociation of Impurities on Copper Current Collector in Electrolyte for Lithium-Ion Batteries. *Materials* **2018**, *11*, 1256. [CrossRef] [PubMed]
16. Luo, S.; Zhou, L.; Wang, X.; Cao, X.; Nie, X.; He, W. Surface Nanocrystallization and Amorphization of Dual-Phase TC11 Titanium Alloys under Laser Induced Ultrahigh Strain-Rate Plastic Deformation. *Materials* **2018**, *11*, 563. [CrossRef] [PubMed]
17. Xie, S.; Xu, J.; Chen, Y.; Tan, Z.; Nie, R.; Wang, Q.; Zhu, J. Indentation Behavior and Mechanical Properties of Tungsten/Chromium co-Doped Bismuth Titanate Ceramics Sintered at Different Temperatures. *Materials* **2018**, *11*, 503. [CrossRef] [PubMed]
18. Lesiuk, G.; Smolnicki, M.; Rozumek, D.; Krechovska, H.; Student, O.; Correia, J.; Mech, R.; De Jesus, A.M.P. Study of the fatigue crack growth in long-term operated mild steel under mixed-mode (I+II, I+III) loading conditions. *Materials* **2019**, in press.

© 2019 by the authors. Licensee MDPI, Basel, Switzerland. This article is an open access article distributed under the terms and conditions of the Creative Commons Attribution (CC BY) license (http://creativecommons.org/licenses/by/4.0/).

Article

Probabilistic Fatigue/Creep Optimization of Turbine Bladed Disk with Fuzzy Multi-Extremum Response Surface Method

Chun-Yi Zhang [1], Zhe-Shan Yuan [1], Ze Wang [1], Cheng-Wei Fei [2,*] and Cheng Lu [3]

1. School of Mechanical and Power Engineering, Harbin University of Science and Technology, Key Laboratory of Advanced Manufacturing and Intelligent Technology, Ministry of Education, Harbin 150080, China; zhangchunyi@hrbust.edu.cn (C.-Y.Z.); yuanzheshan_ma17@hrbust.edu.cn (Z.-S.Y.); wangze_ma17@hrbust.edu.cn (Z.W.)
2. Department of Aeronautics and Astronautics, Fudan University, Shanghai 200433, China
3. School of Aeronautics, Northwestern Polytechnical University, Xi'an 710072, China; lucheng2013@163.com
* Correspondence: cwfei@fudan.edu.cn

Received: 6 August 2019; Accepted: 14 October 2019; Published: 15 October 2019

Abstract: To effectively perform the probabilistic fatigue/creep coupling optimization of a turbine bladed disk, this paper develops the fuzzy multi-extremum response surface method (FMERSM) for the comprehensive probabilistic optimization of multi-failure/multi-component structures, which absorbs the ideas of the extremum response surface method, hierarchical strategy, and fuzzy theory. We studied the approaches of FMERSM modeling and fatigue/creep damage evaluation of turbine bladed disks, and gave the procedure for the fuzzy probabilistic fatigue/creep optimization of a multi-component structure with FMERSM. The probabilistic fatigue/creep coupling optimization of turbine bladed disks was implemented by regarding the rotor speed, temperature, and density as optimization parameters; the creep stress, creep strain, fatigue damage, and creep damage as optimization objectives; and the reliability and GH4133B fatigue/creep damages as constraint functions. The results show that gas temperature T and rotor speed ω are the key parameters that should be controlled in bladed disk optimization, and respectively reduce by 85 K and 113 rad/s after optimization, which is promising to extend bladed disk life and decrease failure damages. The simulation results show that this method has a higher modeling accuracy and computational efficiency than the Monte Carlo method (MCM). The efforts of this study provide a new useful method for overall probabilistic multi-failure optimization and enrich mechanical reliability theory.

Keywords: fuzzy theory; multi-extremum response surface method; bladed disk; fatigue creep; probabilistic optimization

---

1. Introduction

Mechanical structures are usually assembled by a several components; for example, the rotor system of an aero engine is assembled by a spindle, disk, blade, and other components [1]. If we directly establish the reliability optimization design model of an overall structure involving multi-material, multi-disciplinary, and multi-physics structures, the computational burden will become very large in analysis, so that computational efficiency is unacceptable [2]. Therefore, it is significant to propose an efficient method for an overall reliability optimization design of multi-component and multi-failure modes, to make computational precision and efficiency satisfy engineering requirements.

Recently, numerous methods on structural reliability optimization design have emerged [3–5]. The response surface method (RSM) is widely used in reliability optimal design for high efficiency and precision. Zhang et al. [6] firstly proposed an extremum response surface method to complete the reliability optimization of a two-link flexible manipulator; Fei et al. [7–9] studied an importance degree model with the extremum response surface method for the dynamic reliability optimization design of a mechanical assembly relationship such as turbine blade-tip radial clearance. However, the traditional RSM can't meet the reliability optimization design of complex mechanical structures in modeling accuracy and simulation efficiency. To solve this issue, advanced response surface methods were developed recently. Song et al. [10] established a multiple response surface model by using back propagation-artificial neural network to construct a limit state function and establish a multi-objective reliability-based optimization model with a dynamic multi-objective particle swarm optimization algorithm for a reliability optimization design of an aero-engine blisk under fluid–thermal–structure coupling. Hamzaoui et al. [11] proposed an integrated method for estimating the resonance stress of blades with super high strength by combining the inverse of artificial neural network inverse (ANNI) with the Nelder–Mead optimization method. Rodríguez et al. [12] applied a probabilistic design procedure to a group of 10 blades of a low pressure (LP) stage steam turbine of 110 MW, in order to compute the stress changes and reliability due to variations in: damping, natural frequencies, vibration magnitude, and density. The computed vibration stresses were analyzed by applying probability distributions and statistical parameters of input and output to compute the useful life. Wang et al. [13] introduced evidence variables and fuzzy variables to describe cognitive uncertainty parameters and presented a novel dual-stage reliability analysis framework where the first stage incorporates the evidence information by the belief and plausibility measures and the second stage incorporates the fuzzy information by a membership function-like formula. Gao et al. [14] proposed an accurate and efficient fatigue prognosis based on a distributed collaborative response surface method, a substructure-based distributed collaborative probabilistic analysis method (SDCPAM), and a substructure analysis method. Ai et al. [15] discussed a probabilistic framework for fatigue reliability analysis. These works implement reliability-based optimization for many analytical objectives, through analyzing the submodels and then processing the response of submodels to carry out the overall design and analysis. The basic thought in the above works for handling multi-objective design problems provides an enlightened insight to reveal the overall reliability-based optimization design of turbine bladed disks with many failure modes, such as stress failure, strain failure, fatigue damage, creep damage, and so forth. However, since scientific research has its own development laws, the reliability optimization design was carried out in one failure mode at that time, without considering the correlation between the failure modes, and the fuzziness of the constraint boundary conditions.

Most works on aero-engine turbine blades regard the randomness of variables (parameters). Alongside the randomness, actually, some parameters in blade models, such as density, temperature, elastic modulus, boundary conditions, and so forth, possess obvious fuzziness centering on a certain value [16]. In fact, the probabilistic fatigue/creep optimization design of turbine bladed disks involves an obvious fuzziness for design parameters and constraint conditions as well as the coupling among many failure modes such as stress failure, strain failure, creep damage, fatigue damage, and so on [15–18]. Meanwhile, the fuzziness and coupling seriously negatively influence the design precision and efficiency of multi-object optimization when the above methods are directly applied. Therefore, it is urgent to propose an effective method for multi-object reliability-based optimization, in which the fuzziness for design parameters and constraint conditions as well as the coupling among many failure modes are fully considered in order to improve the modeling accuracy and simulation efficiency.

The objective of this paper is to attempt to propose a fuzzy multi-extremum response surface method (FMERSM) regarding failure correlation and parameter fuzziness, to improve the accuracy and efficiency of the overall dynamic reliability optimization design for a multi-component structure with multi-failure mode, by reasonably handling the transients. Then, the probabilistic fatigue/creep optimization design of an aero-engine bladed disk was effectively implemented with respect to this method, and the developed FMERSM is validated by a comparison of methods.

The remainder of this paper is organized as follows. The fuzzy multi-extremum response surface method (FMERSM) is studied in Section 2, comprising the FMERSM modeling approach, fatigue/creep theory, and the basic thought of the comprehensive probabilistic optimization of a bladed disk with FMERSM. Section 3 implements the fuzzy reliability-based optimization of bladed disk fatigue/creep damage including a parameters selection, finite element (FE) modeling, surrogate modeling, probabilistic fatigue/creep analysis, and method validation. In Section 4, some main conclusions are summarized.

2. Methods and Models

The extremum response surface method (ERSM) was firstly developed to simplify the modeling complexity for the transient probabilistic design of mechanical structures by considering the extreme values of the response process in sample extraction [6]. ERSM has been validated to have high-computational efficiency and acceptable accuracy relative to RSM, in the probabilistic design and optimization of aerospace structures/components [7,8,19–22]. The multi-extremum response surface method was proposed to handle the multi-model problem in the transient probabilistic analysis of multi-component structures, multi-discipline, and multi-failure modes by assimilating ERSM [19,23–25]. In most of the structural probabilistic designs, in fact, influential parameters and constraint conditions hold obvious fuzziness and seriously influence design precision. Therefore, it is reasonable to consider the fuzziness of design parameters and constraint conditions to improve the probabilistic design of structures, especially with multi-failure modes or multi-component structures. In respect of the heuristic thought of MERSM, this paper develops FMERSM with the consideration of fuzzy parameters and constraints to implement the fuzzy reliability-based optimization of bladed disk fatigue/creep damage.

2.1. FMERSM Modeling

Assuming that a structure system includes m components and one component has n failure modes $(m,n \in \mathbf{Z})$ (the sample number of failure models is assumed in this study), as well as X^{ij} indicating the input random variables of the jth failure mode in the ith component (for instance, the creep failure of a blade in a bladed disk system) and $y^{(ij)}(t, X^{(ij)})$ is the corresponding output response, enough of a data set $\{y^{ij}{}_{max}(t, X^{(ij)}): j \in \mathbf{Z}_+\}$ consisting of the maximum output responses of $y^{(ij)}(t, X^{(ij)})$ in the time domain is employed to fit the extremum output response y [24]:

$$y = f(X) = \left\{ y^{(ij)}_{max}\left(X^{(ij)}\right) \right\}_{i=1,2,\cdots,m;j=1,2,\cdots,n} \quad (1)$$

When the quadratic polynomials are considered, Equation (1) is rewritten as:

$$y = a_0 + BX + X^T CX \quad (2)$$

Regarding the fuzziness and randomness of data in Equation (2), the model comprising numerous sub-models ($\widetilde{y}^{(11)}_{max}, \widetilde{y}^{(12)}_{max}, \cdots, \widetilde{y}^{(i1)}_{max}, \widetilde{y}^{(i2)}_{max}, \cdots, \widetilde{y}^{(ij)}_{max}$), the FMERSM model, for multi-failure structure, can be structured as:

$$\begin{cases} \widetilde{y}^{(11)}_{max} = f\left(X^{(11)}\right) = \widetilde{A}^{(11)}_0 + \widetilde{B}^{(11)} X^{(11)} + \left(X^{(11)}\right)^T \widetilde{C}^{(11)} X^{(11)} \\ \widetilde{y}^{(12)}_{max} = f\left(X^{(12)}\right) = \widetilde{A}^{(12)}_0 + \widetilde{B}^{(12)} X^{(12)} + \left(X^{(12)}\right)^T \widetilde{C}^{(12)} X^{(12)} \\ \vdots \\ \widetilde{y}^{(ij)}_{max} = f\left(X^{(ij)}\right) = \widetilde{A}^{(ij)}_0 + \widetilde{B}^{(ij)} X^{(ij)} + \left(X^{(ij)}\right)^T \widetilde{C}^{(ij)} X^{(ij)} \end{cases} \quad (3)$$

in which $\widetilde{X}^{(ij)}$ is the fuzzy random input variable vector of the jth failure mode in the ith component, and $\widetilde{y}^{(ij)}_{max}$ is the corresponding extremum output response. $\widetilde{A}^{(ij)}_0, \widetilde{B}^{(ij)}$ and $\widetilde{C}^{(ij)}$ are the constant term, linear term, and quadratic term of the jth failure mode in the ith component, respectively. $\widetilde{B}^{(ij)}$, $\widetilde{C}^{(ij)}$ and $\widetilde{X}^{(ij)}$ are denoted by:

$$\widetilde{B}^{(ij)} = \left[b_1^{ij}, b_2^{ij}, \cdots, b_k^{ij}\right] \tag{4}$$

$$\widetilde{C}^{(ij)} = \begin{pmatrix} c_{11}^{(ij)} & \cdots & 0 \\ \vdots & \ddots & \vdots \\ c_{k1}^{(ij)} & \cdots & c_{kk}^{(ij)} \end{pmatrix} \tag{5}$$

$$\widetilde{X}^{(ij)} = \left[X_1^{(ij)}, X_2^{(ij)}, \cdots, X_k^{(ij)}\right]^T \tag{6}$$

where b_m^{ij}, $c_{mn}^{(ij)}$, $X_m^{(ij)}$ ($m, n = 1, 2, \ldots, k$) are elements (or components) in $\widetilde{B}^{(ij)}$, $\widetilde{C}^{(ij)}$ and $\widetilde{X}^{(ij)}$ respectively.

The modeling process of Equation (3) regards the randomness and fuzziness of design parameters and constraints based on FMERSM. Therefore, this model (Equation (3)) is called a FMERSM model in this paper.

2.2. Fatigue/Creep Modeling for Probabilistic Optimization of Bladed Disks

Under fatigue/creep coupling failure mode, this paper adopts FMERSM to complete the fuzzy probabilistic fatigue/creep optimization of bladed disks. For a structure system with m components, $\widetilde{x}_i$ indicates the fuzzy optimization parameters of the ith component. The main plan is to minimize the objective function $f(\widetilde{x}_1, \widetilde{x}_2, \cdots, \widetilde{x}_n)$ subject to the overall reliability performance $R(x, w, D_c, D_f)$ and coupling critical damage D_{cr}, which is a single-objective constrained optimization problem. The sub-plan is to maximize the reliability R_i ($R_i = R\left(R_i^{(1)}, R_i^{(2)}, \cdots, R_i^{(k)}\right)$) of the ith component subject to mechanical load and constraints, which is a multi-objective constrained optimization problem. By introducing pseudo-variables [26], the cyclic optimization between the main plan and sub-plans is done until the convergence condition is satisfied. The fuzzy probabilistic optimization model is shown in Equation (7).

$$\begin{cases} \text{find } \widetilde{x} = (x_1, x_2, \cdots, x_n)^T \\ \min\ f(\widetilde{x}_1, \widetilde{x}_2, \cdots \widetilde{x}_n) = E\left\{\sum_{i=1}^{l} f_i(\widetilde{x}_i)\right\} \\ \text{subject to} \begin{cases} R(x, \omega, D_c, D_f) = R(R_1, R_2 \cdots R_m) \geq R_0 \\ D_c + D_f \leq D_{cr} \end{cases} \end{cases} \xrightarrow[R_i]{\widetilde{x}_i} \begin{cases} \text{find } \widetilde{x}_i = [x_{i1}, x_{i2}, \cdots, x_{in}] \\ \max\ R_i = R\left(R_i^{(1)}, R_i^{(2)} \cdots R_i^{(k)}\right) \\ \text{subject to} \begin{cases} \widetilde{g}_j(x_j) \subseteq \widetilde{G}_j \\ x_{i1}^L \leq x_i \leq x_{in}^U \end{cases} \end{cases} \tag{7}$$

where $\widetilde{x}_i$ is the ith design variable; and w is the random parameters of mechanical load and material property. x_{i1}^L, x_{in}^U represent the lower and upper limit of the ith fuzzy design variables; D_c is the total amount of creep damages; D_f is the total amount of fatigue damages; D_{cr} is fatigue–creep coupled critical damage; $\widetilde{g}_j(x_i)$ denotes the stress and deformation of a component; and $\widetilde{G}_j$ is the allowable range of $\widetilde{g}_j(x_i)$. By the λ level-cut method, the fuzzy subset $\widetilde{G}_j$ is decomposed into the common set $G_j(\lambda^*)$, as explained in Equation (8); then, the problem of fuzzy probabilistic constrained optimization can be transformed into the conventional probabilistic optimization design problem [27].

$$G_j(\lambda^*) = \left\{g \middle| u_{\widetilde{G}_j}(g) \geq \lambda^*, j = 1, 2, \cdots, J\right\} \tag{8}$$

where $u_{\widetilde{G}_j}(g)$ is allowable constraint of the jth component stress and deformation; and λ^* is optimal horizontal cut set.

2.3. Miner Linear Accumulation Damage Law

Under the interaction between fatigue and creep, the overall damage of the structure is equal to the sum of fatigue damage and creep damage, which is the Miner linear cumulative damage law [28] as follows:

$$\begin{cases} D_f + D_c \leq D_{cr} \\ D_f = \sum_{j=1}^{nf} \dfrac{n_j}{N_{jf}} \\ D_c = \sum_{i=1}^{nc} \dfrac{t_i}{T_{ic}} \end{cases} \quad (9)$$

in which n_f is the number of stresses acting on a component; n_j is the number of cycles acted by the jth stress; N_{jf} is the fatigue life under the jth acting stress; n_c is the number of stress levels; t_i is the hold time of the ith stress; and T_{ic} is the creep failure time of the ith stress.

When the structure is destroyed ($D_{cr} = 1$), the relationship between D_f and D_c [29] is:

$$D_f = F(D_c) = 2 - e^{\theta_1 D_c} + \dfrac{e^{\theta_1} - 2}{e^{-\theta_2} - 1}\left(e^{-\theta_2 D_c} - 1\right) \quad (10)$$

where θ_1 and θ_2 are fatigue–creep characteristic parameters.

The strain fatigue life prediction model is used to predict the low-cycle fatigue life.

$$\dfrac{\Delta \varepsilon}{2} = \dfrac{\sigma_f}{E}\left(2N_f\right)^b + \varepsilon_f\left(2N_f\right)^c \quad (11)$$

in which $\Delta \varepsilon$ is the amplitude of total strain; N_f is the fatigue life; σ_f is the fatigue strength coefficient; ε_f is the fatigue ductility coefficient; b is the fatigue strength index; and c is the fatigue ductility index.

The creep life prediction equations commonly used in material manuals include creep life prediction equations and thermal strength parameter synthesis equations. The persistence equation is expressed in the form of the thermal intensity parameter synthesis equation.

$$\lg \sigma = a_0 + a_1 p + a_2 p^2 + a_3 p^3 \quad (12)$$

$$p = (\lg t_i + c)C \;(i = 0, 1, 2, 3) \quad (13)$$

in which σ is durable strength; $a_r(r = 0, 1, 2, 3)$ is the undetermined coefficient in which r indicates the subscript of the rth coefficient in Equation (12); p is the thermal intensity parameter; t_i is the hold time of the ith stress; and c and C are the constants related to fatigue ductility and temperature, respectively, which were generally gained by experiments.

2.4. Basic Thought of Probabilistic Fatigue/Creep Optimization with FMERSM

The basic thought of probabilistic fatigue/creep optimization with FMERSM is illustrated below. (1) Regard material density, gas temperature, pneumatic pressure, elastic modulus, and thermal expansion coefficient as input variables, and the maximum creep stress, maximum creep strain, maximum fatigue damage, and maximum creep damage as output responses. (2) Carry out the deterministic analysis of a bladed disk based on FE models with the consideration of design parameters. (3) Obtain the fatigue damage and creep damage of a bladed disk under each load by the fatigue–creep damage equation of GH4133B discussed in Section 2.2 and Miner linear damage accumulation law introduced in Section 2.3. (4) Considering the randomness and fuzziness of input variables, enough samples of input random variables are extracted by the Latin hypercube sampling technique [30]. (5) Calculate the dynamic responses of bladed disk creep stress, creep strain, creep damage, and fatigue damage in the time domain for all input samples by FE models, and extract the maximum values of dynamic output responses as new output responses to establish the FMERSM function. (6) Perform a probabilistic analysis of a bladed disk based on the FMERSM function. (7) Complete the probabilistic fatigue/creep optimization of a turbine bladed disk by the fuzzy probabilistic optimization model with FMERSM and decoupling coordination iterative solution. The flowchart is shown in Figure 1.

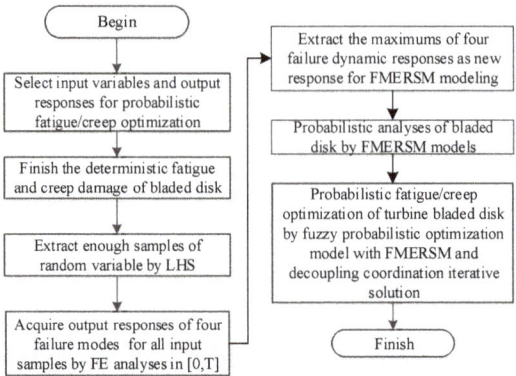

Figure 1. Flow chart of reliability optimization based on the fuzzy multi-extremum response surface method (FMERSM) method.

3. Fuzzy Probabilistic Fatigue/Creep Optimization of Turbine Bladed Disk

In this section, the fuzzy probabilistic fatigue/creep optimization of a turbine bladed disk is performed with respect to the proposed FMERSM and established probabilistic optimization model in Section 2.

3.1. Parameters Preparation

With respect to the material test, the fatigue/creep material parameters θ_1 and θ_2 of an aero-engine turbine bladed disk with a GH4133B superalloy at the temperature of 600 °C and experimental load of 18 KN is 0.36 and 6.5, respectively. The fatigue–creep damage curve of GH4133B superalloy (Ni–Cr-based precipitation hardening-type deformation high-temperature alloy) is shown in Figure 2. In this study, we selected a 1/40 turbine bladed disk of an aero-engine as the object of study, and a GH4133B superalloy as the material of the bladed disk [31]. Density ρ, rotational speed ω, temperature T, pneumatic pressure p, elastic modulus E, and thermal expansion coefficient α are considered as fuzzy variables. Moreover, in respect of engineering practice, the length of the fuzzy region is defined as 0.05 times the mean value, as shown in Table 1. The parameters in Table 1 are assumed to obey normal distribution, and are mutually independent.

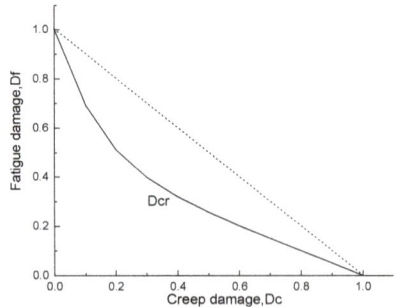

Figure 2. Curves of GH4133B fatigue–creep damage.

Table 1. Distribution characteristics of input random variables.

Random Variables	Mean	Length of Fuzzy Zone	Distribution
Density, ρ, kg·m^{-3}	8210	410.5	Normal
Rotor speed, ω, rad·s^{-1}	1168	58.4	Normal
Temperature T, K	873.15	43.658	Normal
Pneumatic pressure p, MPa	0.1	0.005	Normal
Elastic modulus E, MPa	163,000	8150	Normal
Thermal expansion coefficient α, $\times 10^{-6}$ °C^{-1}	9.4	0.47	Normal

3.2. Deterministic Analysis of Bladed Disk

The finite element (FE) models of the blade and disk are shown in Figures 3 and 4. The FE model of the blade consists of tetrahedrons with 39,547 elements, and the FE model of the disk consists of tetrahedrons with 58,271 elements. The number of cells (meshes) is required by the convergence analyses of the responses [15,32,33]. The FE basic equations of the bladed disk comprising a shape function of the tetrahedron in Equation (14) [34], geometric equation in Equation (15) [35], physical equation in Equation (16) [36], and Norton implicit creep equation in Equation (17) [37] are analyzed with regard to the means of the parameters in Table 1. From this analysis, the distributions of the creep stress and creep strain of the bladed disk are drawn in Figures 5–8. As seen in Figures 5–8, the maximum creep stress and maximum creep strain of the bladed disk are at the blade-root and disk tenon groove tip, respectively.

$$N_i = \frac{1}{6v}(a_i + b_i x + c_i y + d_i z), \; i = 1, 2, 3, 4 \tag{14}$$

$$\begin{cases} \varepsilon_x = \frac{\partial u}{\partial x}, \varepsilon_y = \frac{\partial v}{\partial y}, \varepsilon_z = \frac{\partial w}{\partial z} \\ \gamma_{xy} = \frac{\partial u}{\partial y} + \frac{\partial v}{\partial x}, \gamma_{yz} = \frac{\partial v}{\partial z} + \frac{\partial w}{\partial y}, \gamma_{zx} = \frac{\partial w}{\partial x} + \frac{\partial u}{\partial z} \end{cases} \tag{15}$$

$$\{\sigma\} = [D]\{\varepsilon\} \tag{16}$$

$$\varepsilon_c = c_1 \sigma^{C_2} e^{-C_3/T} \tag{17}$$

where v is the volume of the tetrahedron; a_i, b_i, c_i and d_i are the related coefficients of node geometry; ε_x, ε_y, ε_z and γ_{xy}, γ_{yz}, γ_{zx} are the elastic line strains and shear strains along the x, y and z directions, respectively; $[\sigma] = [\sigma_x, \sigma_y, \sigma_z, \tau_{xy}, \tau_{yz}, \tau_{zx}]$ (σ-structural stress) and $[\varepsilon] = [\varepsilon_x, \varepsilon_y, \varepsilon_z, \gamma_{xy}, \gamma_{yz}, \gamma_{zx}]$ are the stress and strain of the components; $[D]$ is the elastic matrix; c_1, c_2, and c_3 stand for experimental coefficients.

Figure 3. Finite element model of the blade.

Figure 4. Finite element model of the disk.

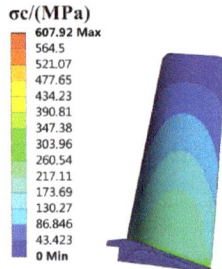

Figure 5. Distribution of blade creep stress.

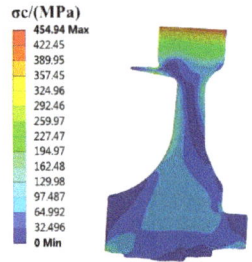

Figure 6. Distribution of disk creep stress.

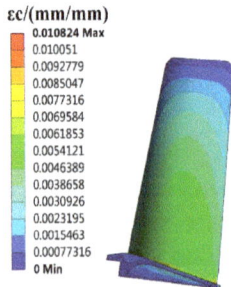

Figure 7. Distribution of blade creep strain.

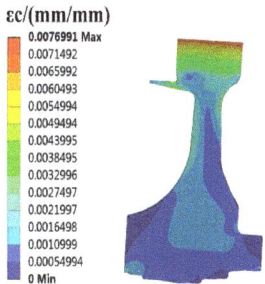

Figure 8. Distribution of disk creep strain.

Based on the low-cycle load spectrum described in [34], the Miner linear cumulative damage law in Equation (9) and fatigue-creep damage relation in Equation (10) of GH4133B [38] were resolved by programming in MATLAB (R2017a) simulation environment. When the number of cyclic loads is 5530, the fatigue damage D_f and creep damage D_c of bladed disk under a cyclic load are shown in Table 2.

Table 2. Results of bladed disk fatigue-creep damage.

	Fatigue Damage D_f	Creep Damage D_c
Blade	0.36363	0.0039
Disk	0.40859	0.0041

3.3. FMERSM Modeling

By the Latin hypercube sampling technique [30], 100 samples on fuzzy input random variables were extracted in respect of the max creep stress, max creep strain, max fatigue damage, and max creep damage of the bladed disk, to acquire model parameters and establish the FMERSM model in Equations (18) and (19).

$$\begin{cases} \bar{y}^{(11)} &= -257.3754 - 0.0444\rho - 0.1486\omega + 2.3354 \times 10^3 T + 1.1507p + 1.4334 \times 10^{-4} E - 0.2077a \\ &\quad + 6.5536 \times 10^{-4}\rho^2 + 0.0132\omega^2 + 3.6612 \times 10^5 T^2 + 0.0102p^2 4.2577 \times 10^{-5} E^2 - 4.8668 \times 10^{-4} a^2 \\ \bar{y}^{(12)} &= 0.0099 - 1.6579 \times 10^{-6}\rho - 1.3072 \times 10^{-5}\omega - 0.0599T - 9.8795 \times 10^{-6} p + 1.4733 \times 10^{-9} E - 1.8500 \times 10^{-6} a \\ &\quad + 2.8940 \times 10^{-6}\rho^2 + 14.4810\omega^2 + 7.4616 \times 10^{-7} T^2 + 5.3444 \times 10^{-4} p^2 + 4.4308 \times 10^{-6} E^2 + 3.7239 \times 10^{-7^2} a^2 \\ \bar{y}^{(13)} &= 1.5805 \times 10^{-4} - 1.3288 \times 10^{-8}\rho - 1.0181 \times 10^{-7}\omega - 2.7108 \times 10^{-4} T - 6.7492 \times 10^{-8} p + 4.3108 \times 10^{-12} E - 1.3172 \times 10^{-8} a \\ &\quad + 2.8445 \times 10^4 \rho^2 + 0.0027\omega^2 + 0.1975T^2 - 6.5354p^2 - 2.6096E^2 + 0.0043a^2 \\ \bar{y}^{(14)} &= 3.0155 \times 10^{-4} + 3.2551 \times 10^{-8}\rho - 1.0601 \times 10^{-7}\omega + 0.00016T - 7.4068 \times 10^{-8} p + 5.8844 \times 10^{-12} E - 1.2685 \times 10^{-7} a \\ &\quad + 2.9205 \times 10^{-5}\rho^2 + 6.7736 \times 10^{-6}\omega^2 + 1.2254 \times 10^{-4} T^2 + 1.8214 \times 10^{-4} p^2 - 8.1228 \times 10^{-7} E^2 + 6.6609 \times 10^{-8} a^2 \end{cases} \quad (18)$$

$$\begin{cases} \bar{y}^{(21)} &= 4.6236 \times 10^3 - 0.0477\rho - 0.3303\omega - 2.4545 \times 10^3 T - 9.4539p + 1.9172 \times 10^{-4} E - 0.2481a \\ &\quad - 23.2706\rho^2 + 0.0031\omega^2 + 11.0768T^2 + 5.9537 \times 10^{-5} p^2 + 6.8958 \times 10^{-4} E^2 + 3.8867 \times 10^{-4} a^2 \\ \bar{y}^{(22)} &= 0.1404 + 1.4686 \times 10^{-7}\rho - 2.4194 \times 10^{-5}\omega + 0.1010T - 3.1221 \times 10^{-4} p + 2.1079 \times 10^{-9} E - 1.5086 \times 10^{-5} a \\ &\quad + 2.7152 \times 10^{-11}\rho^2 - 1.1026 \times 10^{-11}\omega^2 + 5.5067 \times 10^{-9} T^2 + 2.0951 \times 10^{-9} p^2 - 8.5852 \times 10^{-12} E^2 - 4.0029 \times 10^{-11} a^2 \\ \bar{y}^{(23)} &= -8.7299 \times 10^{-5} - 2.6642 \times 10^{-9}\rho - 5.0414 \times 10^{-8}\omega - 1.9935 \times 10^{-4} T + 3.5076 \times 10^{-7} p + 7.8942 \times 10^{-12} E - 3.7701 \times 10^{-8} a \\ &\quad + 4.0029 \times 10^{-4}\rho^2 + 3.1495 \times 10^{-6}\omega^2 + 7.2796 \times 10^{-5} T^2 + 6.8958 \times 10^{-4} p^2 - 9.4331 \times 10^{-5} E^2 + 3.8867 \times 10^{-4} a^2 \\ \bar{y}^{(24)} &= 0.0059 + 6.2600 \times 10^{-8}\rho - 1.0922 \times 10^{-6}\omega + 0.0140T - 4.3691 \times 10^{-6} p + 3.2447 \times 10^{-11} E + 1.8682 \times 10^{-6} a \\ &\quad + 1.9739 \times 10^{-9}\rho^2 + 3.6047 \times 10^{-12}\omega^2 + 3.3291 \times 10^{-10} T^2 + 0.0046p^2 + 0.0048E^2 - 7.3346 \times 10^{-4} a^2 \end{cases} \quad (19)$$

The FMERSM models in Equations (18) and (19) are used to perform the probabilistic fatigue/creep optimization of a bladed disk involving sensitivity analysis and reliability analysis in the following subsection.

3.4. Probabilistic Fatigue/Creep Optimization of Bladed Disk

Regarding the FMERSM models in Equations (18) and (19), the reliability sensitivity index of input random variables for a bladed disk was obtained in Table 3 and Figure 9 by sensitivity analysis with MC simulation.

Table 3. Sensitivity index of a bladed disk.

	Blade			Disk	
Variables	Sensitivity	Effect Probability %	Variable	Sensitivity	Effect Probability %
ρ	0.09855	10.55	ρ	0.22067	24.04
ω	0.477722	51.14	ω	0.386637	40.99
T	0.241222	25.82	T	0.252026	26.72
p	0.073895	7.91	p	−0.0121	1.28
E	0.032676	3	E	0.02191	2.32
α	−0.01	1.07	α	0.043787	4.64

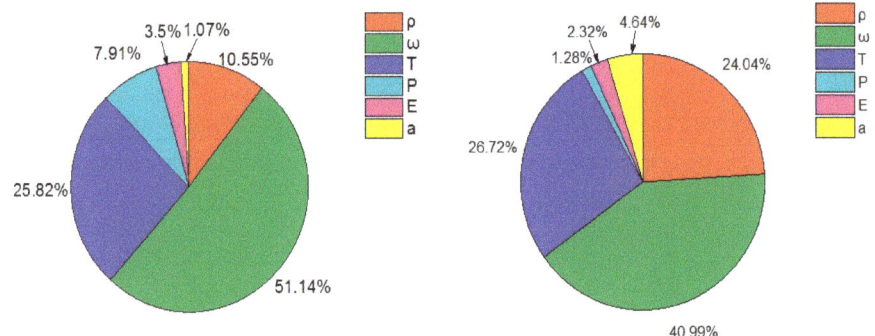

Figure 9. Sensitivity indexes of parameters on bladed disk coupling failure.

As shown in the sensitivity analysis of a bladed disk, rotor speed w and temperature T are the main factors and greatly influence the coupling failure of a bladed disk as the two largest sensitivity degrees and effect probabilities, while the other parameters have little impact on the coupling failure of a bladed disk as smaller sensitivity degrees and effect probabilities. We extract high-sensitivity input random variables as design variables to conduct the fuzzy probabilistic fatigue/creep optimization of a bladed disk. In the main planning model, the reliability product of $R_1 \cdot R_2$ and the coupling critical damage D_{cr} for the bladed disk are taken as the constraints. In the sub-planning model, the parameters (ω,T) with a high-sensitivity index are regarded as the design variables. The creep stress σ_c, creep strain ε_c, fatigue damage D_f, creep damage D_c, maximum blade reliability R_1, and maximum disk reliability R_2 were evaluated. The allowable comprehensive reliability of the bladed disk is $R_0 = 0.99$. The optimal level λ^* is solved by the fuzzy comprehensive evaluation method [13,23,39], and the substitution of λ^* into the asymmetric fuzzy optimized conversion condition (Equation (8)). The allowable means of a bladed disk with the corresponding failure modes are shown in Table 4. The fuzzy probabilistic fatigue/creep optimization model of a bladed disk was established as illustrated in Figure 10, and the optimization models were solved by the MATLAB program and iteratively solved for all levels. The optimization results are listed in Table 5.

Table 4. Optimal level threshold and allowable mean of a bladed disk.

Blade				Disk			
Optimal Level Threshold		Allowable Mean		Optimal Level Threshold		Allowable Mean	
$\lambda^*_{\sigma c1}$	0.3558	$\widetilde{\sigma}_{c10}$	677.92	$\lambda^*_{\sigma c2}$	0.6008	$\widetilde{\sigma}_{c20}$	654.94
$\lambda^*_{\varepsilon c1}$	0.8051	$\widetilde{\varepsilon}_{c10}$	2.010824	$\lambda^*_{\varepsilon c2}$	0.8516	$\widetilde{\varepsilon}_{c20}$	1.0076991
λ^*_{Dc1}	0.6720	$\widetilde{D}_{c10}$	0.2039	λ^*_{Dc2}	0.1608	$\widetilde{D}_{c20}$	0.2041
λ^*_{Df1}	0.0260	$\widetilde{D}_{f10}$	0.96363	λ^*_{Df2}	0.8392	$\widetilde{D}_{f20}$	0.90895

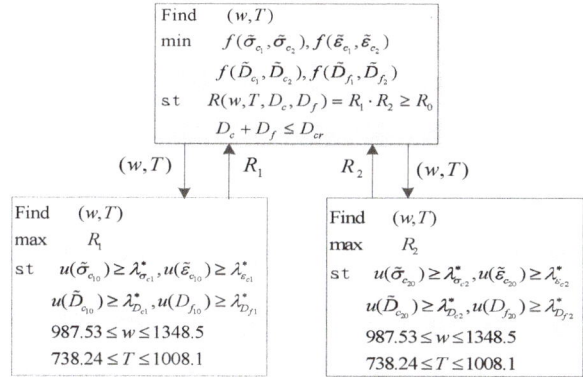

Figure 10. Fuzzy probabilistic fatigue/creep optimization model of a bladed disk.

Table 5. Optimized results of a bladed disk.

Design Variable	Original Data	Optimization Results
ω, rad·s^{-1}	1168	1055.1
T, K	873.15	788.15

3.5. FMERSM Validation

To verify the effectiveness of FMERSM, the reliability-based optimization of a bladed disk was completed with MCM and FMERSM, based on the same variables in Table 1 and computing conditions. For dynamic probabilistic analyses under different simulations (10^2, 10^3, 10^4, and 10^5), the computing time and reliability degrees of a bladed disk are listed in Table 6. The optimization results of object functions under different simulations are listed in Table 7.

Table 6. Dynamic probabilistic computational results with different methods. FMERSM: fuzzy multi-extremum response surface method, MC: Monte Carlo.

Number of Samples	Computational Time, s		Reliability Degree %		Precision of FMERSM
	MC Method	FMERSM	MC Method	FMERSM	
10^2	32400	0.203	99	98.6	0.996
10^3	72000	0.279	99.7	99.5	0.998
10^4	432000	0.437	99.83	99.60	0.9977
10^5	-	4.43	-	99.962	-

Table 7. Results of bladed disk optimization design with different methods.

Objective Function	Before Optimization	MCM		FMERSM	
		After Optimization	Reduction	After Optimization	Reduction
σ_{c_1}, MPa	607.92	548.66	9.8%	487.08	19.9%
σ_{c_2}, MPa	454.94	423.03	7%	368.8	18.93%
ε_{c_1}, m/m	0.010824	0.001298	88%	0.0073988	31.64%
ε_{c_2}, m/m	0.0076991	0.0076629	0.47%	0.0065652	14.77%
Df_1	0.36363	0.30452	16.25%	0.24822	31.74%
Df_2	0.40859	0.39375	4.52%	0.29315	28.3%
D_{c_1}	0.0039	0.0036	7.69%	0.0025	35.9%
D_{c_2}	0.0041	0.00409	0.24%	0.0032	21.95%
R	95	99.515	-	99.635	-

As shown in Table 6, the following conclusions were obtained from the probabilistic failure analysis of a bladed disk. (1) The MC method does not have computing time at 10^5 simulations, because the MC method cannot perform the calculation for a too-large computational burden for a probabilistic analysis of bladed disk FE models. Thus, it is inefficient for the MC method to conduct the design analysis of a complex structure with large-scale simulations. (2) The time-cost for the probabilistic analysis of a bladed disk increases with the increase of MC simulations. (3) The time consumption of the FMERSM is far less than that of the MC method for the same number of simulations. For instance, the FMERSM only spends 0.437 s for 10,000 simulations, which is only about $1/10^6$ that of the MC method. Meanwhile, the strength of the FMERSM in time computation is more obvious with increasing simulations. Thus, it is demonstrated that the efficiency of the FMERSM is far higher than that of the MC method in calculation, and the FMERSM is an efficient approach replacing FE models for the probabilistic analysis of a complex structure with many components or multi-failure modes. (4) For the same simulations, the reliability degrees of bladed disk coupling dynamic failure probability with FMERSM are almost consistent with those of the MC method. Moreover, the reliability degree of the bladed disk increases and becomes higher with the rise of simulations. It is illustrated that more precise results such as the reliability degree can be gained by increasing the number of MC simulations against the response surface models, for structure design analysis from a probabilistic perspective.

As revealed in Table 7, summarized from the probabilistic fatigue/creep optimization of a bladed disk, the creep stress, creep strain, fatigue damage, and creep damage of the blade in respect of FMERSM are reduced by 19.9%, 18.93%, 31.64%, and 14.77%, respectively. Meanwhile, the MC method reduces the creep stress, creep strain, fatigue damage, and creep damage of the disk by 9.8%, 7%, 88%, and 0.47%, respectively. The comprehensive reliability index of the bladed disk was increased from 99.515% to 99.635%. It is verified that the FMERSM is workable for the fuzzy probabilistic fatigue/creep optimization of complex structures, similar to a turbine bladed disk.

In summary, the developed FMERSM has high modeling precision and simulation efficiency for the comprehensive reliability optimization design for multi-component structures with multi-failure modes.

4. Conclusions

The objective of this study is to develop a high-efficient reliability-based optimization method, called the fuzzy multi-extremum response surface method (FMERSM), for the probabilistic fatigue/creep coupling optimization of a turbine bladed disk. This paper has investigated the theory and modeling of FMERSM, and gives the procedure of probabilistic optimization of a multi-component structure with multi-failure modes for the fuzzy probabilistic fatigue/creep optimization of a turbine bladed disk with the considerations of the correlation of the failure modes and the fuzziness of the constraint boundary conditions. Through the works in this study, some conclusions are summarized as follows:

(1) With regard to the probabilistic failure analysis of a bladed disk, we find that the FMERSM costs less analytical time (0.437 s for 10,000 simulations), and thus has high computational efficiency relative to the Monte Carlo (MC) method (432,000 s for 10,000 simulations), but has an acceptable computational precision (99.77%) of the reliability degree, which is almost consistent with the FE method based on MC simulation with a reliability degree of 0.9983. Moreover, the strengths of the proposed FMERSM in modeling and simulation become more obvious with the increase of simulations.

(2) In terms of the probabilistic fatigue/creep optimization of a bladed disk, it is illustrated that the developed FMERSM is more workable than the MC method. The reason is that the optimal parameters, including design parameters and optimization objects, are preferable by larger reductions (19.9%, 18.93%, 31.64%, and 14.77% for the creep stress, creep strain, fatigue damage, and creep damage of the blade, respectively), and a higher reliability degree of 99.635%.

The efforts of this paper provide a useful way for high-precise modeling and high-efficient simulation for the fuzzy comprehensive probabilistic optimization of multi-failure/multi-component structures, because the accuracy of the model is close to that of the MC method, while the calculation time is only $1/10^6$. Meanwhile, this work enriches the theory of mechanical reliability.

Author Contributions: Conceptualization, C.-Y.Z. and C.-W.F.; methodology, Z.-S.Y.; software, Z.W.; validation, Z.-S.Y., C.L., and C.-W.F.; formal analysis, C.L.; investigation, Z.W.; resources, C.Z.; data curation, C.F.; writing—original draft preparation, C.-Y.Z. and Z.-S.Y.; writing—review and editing, C.-W.F.; visualization, Z.W.; supervision, C.-W.F. and C.-Y.Z.; project administration, C.-Y.Z.; funding acquisition, C.-Y.Z. and C.-W.F.

Funding: This research was funded by the National Natural Science Foundation of China (Grant No. 51275138 and 51975124), Start-up Research Funding of Fudan University (Grant No. FDU38341) and Excellent Doctorate Cultivating Foundation of Northwestern Polytechnical University (Grant No. CX201932). All the authors would like to thank them.

Conflicts of Interest: The authors declare no conflict of interest.

References

1. Zhang, C.Y.; Lu, C.; Fei, C.W.; Jing, H.Z.; Li, C.W. Dynamic probabilistic design technique for multi-component system with multi-failure modes. *J. Cent. South Univ.* **2018**, *25*, 2688–2700. [CrossRef]
2. Fei, C.W.; Choy, Y.S.; Hu, D.Y.; Bai, G.C.; Tang, W.Z. Transient probabilistic analysis for turbine blade-tip radial clearance with multi-component and multi-physics fields based on DCERSM. *Aerosp. Sci. Technol.* **2016**, *50*, 62–70. [CrossRef]
3. Li, X.; Qiu, H.; Chen, Z.; Gao, L.; Shao, X. A local Kriging approximation method using MPP for reliability-based design optimization. *Comput. Struct.* **2016**, *162*, 102–115. [CrossRef]
4. Majumdar, R.; Ghosh, A.; Das, A.K.; Raha, S.; Laha, K.; Das, S.; Abraham, A. *Artificial Weed Colonies with Neighbourhood Crowding Scheme for Multimodal Optimization*; Springer: New York, NY, USA, 2013.
5. Song, L.K.; Bai, G.C.; Fei, C.W.; Tang, W.Z. Multi-failure probabilistic design for turbine bladed disks using neural network regression with distributed collaborative strategy. *Aerosp. Sci. Technol.* **2019**, *92*, 464–477. [CrossRef]
6. Zhang, C.Y.; Bai, G.C. Extremum response surface method of reliability analysis on two-link flexible robot manipulator. *J. Cent. South Univ.* **2012**, *19*, 101–107. [CrossRef]
7. Fei, C.W.; Bai, G.C.; Tang, W.Z.; Choy, Y.S. Transient reliability optimization for turbine disk radial deformation. *J. Cent. South Univ.* **2016**, *23*, 344–352. [CrossRef]
8. Fei, C.W.; Tang, W.Z.; Bai, G.C. Novel method and model for dynamic reliability optimal design of turbine blade deformation. *Aerosp. Sci. Technol.* **2014**, *39*, 588–595. [CrossRef]
9. Fei, C.W.; Tang, W.Z.; Bai, G.C. Study on the theory, method and model for mechanical dynamic assembly reliability optimization. *Proc. IME Part C J. Mech. Eng. Sci.* **2014**, *228*, 3019–3038. [CrossRef]
10. Song, L.K.; Fei, C.W.; Wen, J.; Bai, G.C. Multi-objective reliability-based design optimization approach of complex structure with multi-failure modes. *Aerosp. Sci. Technol.* **2017**, *64*, 52–62. [CrossRef]
11. Hamzaoui, Y.E.; Rodríguez, J.A.; Hernández, J.A.; Salazar, V. Optimization of operating conditions for steam turbine using an artificial neural network inverse. *Appl. Therm. Eng.* **2015**, *75*, 648–657. [CrossRef]
12. Rodríguez, J.A.; Garcia, J.C.; Alonso, E.; Hamzaoui, Y.E.; Rodríguez, J.M.; Urquiza, G. Failure probability estimation of steam turbine blades by enhanced Monte Carlo Method. *Eng. Fail. Anal.* **2015**, *56*, 80–88. [CrossRef]
13. Wang, C.; Matthies, H.G. Epistemic uncertainty-based reliability analysis for engineering system with hybrid evidence and fuzzy variables. *Comput. Methods Appl. Mech. Eng.* **2019**, *355*, 438–455. [CrossRef]
14. Gao, H.F.; Wang, A.J.; Bai, G.C.; Wei, C.M.; Fei, C.W. Substructure-based distributed collaborative probabilistic analysis method for low-cycle fatigue damage assessment of turbine blade–disk. *Aerosp. Sci. Technol.* **2018**, *79*, 636–646. [CrossRef]
15. Ai, Y.; Zhu, S.P.; Liao, D.; Cirreia, J.A.F.O.; Souto, C.; De Jesus, A.M.P.; Keshtegar, B. Probabilistic modeling of fatigue life distribution and size effect of components with random defects. *Int. J. Fatigue* **2019**, *126*, 165–173. [CrossRef]
16. Song, L.K.; Bai, G.C.; Fei, C.W.; Wen, J. Probabilistic LCF life assessment of turbine discs using DC-based wavelet neural network regression. *Int. J. Fatigue* **2019**, *119*, 204–219. [CrossRef]
17. Cano, S.; Rodríguez, J.A.; Rodríguez, J.M.; García, J.C.; Sierra, F.Z.; Casolco, S.R.; Herrera, M. Detection of damage in steam turbine blades caused by low cycle and strain cycling fatigue. *Eng. Fail. Anal.* **2019**, *97*, 579–588. [CrossRef]
18. Fei, C.W.; Lu, C.; Liem, R.P. Decomposed-coordinated surrogate modelling strategy for compound function approximation and a turbine blisk reliability evaluation. *Aerosp. Sci. Technol.* **2019**, 105466. [CrossRef]

19. Zhang, C.Y.; Wei, J.S.; Jing, H.Z.; Fei, C.W.; Tang, W.Z. Reliability analysis of blisk low fatigue life with generalized regression extreme neural network method. *Materials* **2019**, *12*, 1545. [CrossRef]
20. Lu, C.; Feng, Y.W.; Liem, R.P.; Fei, C.W. Improved kriging with extremum response surface method for structural dynamic reliability and sensitivity analyses. *Aerosp. Sci. Technol.* **2018**, *76*, 164–175. [CrossRef]
21. Gao, H.F.; Fei, C.W.; Bai, G.C. Reliability-based low-cycle fatigue life analysis of turbine blade with thermo-structural interaction. *Aerosp. Sci. Technol.* **2016**, *49*, 289–300. [CrossRef]
22. Fei, C.W.; Bai, G.C.; Tian, C. Extremum response surface method for casing radial deformation probabilistic analysis. *J. Aerosp. Inf. Syst.* **2013**, *10*, 47–52.
23. Lu, C.; Feng, Y.W.; Fei, C.W. Weighted regression-based extremum response surface method for structural dynamic fuzzy reliability analysis. *Energies* **2019**, *12*, 1588. [CrossRef]
24. Zhang, C.Y.; Song, L.K.; Fei, C.W.; Lu, C.; Xie, Y.M. Advanced multiple response surface method for reliability sensitivity analysis of turbine blisk with multi-physics coupling. *Chin. J. Aeronaut.* **2016**, *29*, 962–971. [CrossRef]
25. Titli, A.; Lefevre, T.; Richetin, M. Multilevel optimization methods for non-separable problems and application. *Int. J. Syst. Sci.* **1973**, *4*, 865–880. [CrossRef]
26. Kemp, A.W. Convolutions Involving binomial pseudo-variables. *Sankhyā Indian J. Stat. Ser. A* **1979**, *41*, 232–243.
27. Chakraborty, S.; Sam, P.C. Probabilistic safety analysis of structures under hybrid uncertainty. *Int. J. Numer. Methods Eng.* **2007**, *70*, 405–422. [CrossRef]
28. Hwang, W.; Han, K.S. Cumulative damage models and multi-stress fatigue life prediction. *J. Compos. Mater.* **1986**, *20*, 125–153. [CrossRef]
29. Mahler, M.; Özkan, F.; Aktaa, J. ANSYS creep-fatigue assessment tool for EUROFER97 components. *Nucl. Mater. Energy* **2016**, *9*, 535–538. [CrossRef]
30. Zhai, X.; Fei, C.W.; Wang, J.J.; Choy, Y.S. A stochastic model updating strategy-based improved response surface model and advanced Monte Carlo simulation. *Mech. Syst. Signal Process.* **2017**, *82*, 323–338. [CrossRef]
31. Zhao, R.G.; Li, Q.B.; Jiang, Y.Z.; Luo, X.Y.; Liu, Y.F.; Cai, P.; Chen, Y. Research on transition from short to long fatigue crack propagation of GH4133B superalloy used in turbine disk of aeroengine. *Key Eng. Mater.* **2016**, *697*, 664–669. [CrossRef]
32. Liao, D.; Zhu, S.P.; Correia, J.A.F.O.; De Jesus, A.M.P.; Calçada, R. Computational framework for multiaxial fatigue life prediction of compressor discs considering notch effects. *Eng. Fract. Mech.* **2018**, *202*, 423–435. [CrossRef]
33. Zhu, Z.Z.; Feng, Y.W.; Lu, C.; Fei, C.W. Efficient driving plan and validation of aircraft NLG emergency extension system via mixture of reliability models and test bench. *Appl. Sci.* **2019**, *9*, 3578. [CrossRef]
34. Devloo, P.R.B.; Rylo, E.C. Systematic and generic construction of shape functions for p-adaptive meshes of multidimensional finite elements. *Comput. Methods Appl. Mech. Eng.* **2009**, *198*, 1716–1725. [CrossRef]
35. Hashiguchi, K. A basic formulation of elastoplastic constitutive equations. *Mod. Approaches Plast.* **1993**, *1993*, 39–57.
36. Zhang, C.Y.; Lu, C.; Fei, C.W.; Liu, L.J.; Choy, Y.S.; Su, X.G. Multiobject reliability analysis of turbine blisk with multidiscipline under multiphysical field interaction. *Adv. Mater. Sci. Eng.* **2015**, *2015*, 519–520. [CrossRef]
37. Asraff, A.K.; Sunil, S.; Muthukumar, R.; Ramanathan, T.J. Stress analysis & life prediction of a cryogenic rocket engine thrust chamber considering low cycle fatigue, creep and thermal ratchetting. *Trans. Indian Inst. Met.* **2010**, *63*, 601–606.
38. Zhu, S.P.; Liu, Q.; Peng, W.; Zhang, X.C. Computational-experimental approaches for fatigue reliability assessment of turbine bladed disks. *Int. J. Mech. Sci.* **2018**, *142*, 502–517. [CrossRef]
39. Wang, Y.; Li, Y.; Liu, W.; Gao, Y. Assessing operational ocean observing equipment (OOOE) based on the fuzzy comprehensive evaluation method. *Ocean Eng.* **2015**, *107*, 54–59. [CrossRef]

 © 2019 by the authors. Licensee MDPI, Basel, Switzerland. This article is an open access article distributed under the terms and conditions of the Creative Commons Attribution (CC BY) license (http://creativecommons.org/licenses/by/4.0/).

Article

Influence of Polyurea Composite Coating on Selected Mechanical Properties of AISI 304 Steel

Monika Duda [1], Joanna Pach [2] and Grzegorz Lesiuk [1],*

[1] Department of Mechanics, Materials Science and Engineering, Wrocław University of Science and Technology, Smoluchowskiego 25, 50-370 Wrocław, Poland; monika.duda@pwr.edu.pl
[2] Department of Foundry, Polymers and Automation, Wrocław University of Science and Technology, Smoluchowskiego 25, 50-370 Wrocław, Poland; joanna.pach@pwr.edu.pl
* Correspondence: Grzegorz.lesiuk@pwr.edu.pl

Received: 25 July 2019; Accepted: 23 September 2019; Published: 26 September 2019

Abstract: This paper contains experimental results of mechanical testing of the AISI 304 steel with composite coatings. The main goal was to investigate the impact of the applied polyurea composite coating on selected mechanical properties: Adhesion, impact resistance, static behavior, and, finally, fatigue lifetime of notched specimens. In the paper the following configurations of coatings were tested: EP (epoxy resin), EP_GF (epoxy resin + glass fabric), EP_GF_HF (epoxy resin + glass fabric hemp fiber), EP_PUA (epoxy resin + polyurea) resin, EP_GF_PUA (epoxy resin + glass fabric + polyurea) resin, and EP_GF_HF_PUA (epoxy resin + glass fabric + hemp fiber + polyurea) resin. The highest value of force required to break adhesive bonds was observed for the EP_PUA coating, the smallest for the single EP coating. A tendency of polyurea to increase the adhesion of the coating to the base was noticed. The largest area of delamination during the impact test was observed for the EP_GF_HF coating and the smallest for the EP-coated sample. In all tested samples, observed delamination damage during the pull-off test was located between the coating and the metallic base of the sample.

Keywords: AISI 304; polyurea; composite coating; impact resistance; adhesion; delamination; fatigue

---

1. Introduction

Polymer coatings are the subject of a lot of research in a number of publications [1–5]. They are applied as anti-corrosive agents [6–9] as well as an anti-wear agents preventing abrasion, tearing, and scratches [10–13] due to their specific mechanical properties. In those roles, polyurea, polyurethane, and polyurethane–polyurea resins are mainly used.

Polyurea coatings are increasingly popular in recent years [14–16]. They can be applied on metallic, wooden, and concrete surfaces, or even other plastics. These coatings allow for desired decorative properties to be obtained as well as specific mechanical properties. Polymers are used as cover layers for armed vehicles, ballistic shields [17], loading area of vehicles, and as a waterproof layer on concrete surfaces [18,19]. Additionally, they can absorb vibrations and sound waves. In order to strengthen the layers' properties, they were modified with glass fabric and hemp fiber.

The issue regarding the application of coatings is its poor adhesion to a metal surface. Due to this, in industrial conditions the surface is pre-processed by sandblasting and/or by application of an intermediate layer—primer—based on epoxy resin.

According to the practice of the polyurea coating application process, in the investigation presented in this paper the epoxy resin was used as an intermediate layer. The primer was modified in order to improve the impact resistance and vibration absorption by glass fabric and hemp fiber. In order to determine the influence of each constituent of the layer, there were also prepared samples with and without the polyuria layer. In order to eliminate the influence of mechanical treatment on the results of the experiment, the metallic base material was cleansed with acetone; the sand blasting process was not applied.

The coating was modified with natural fibers due to their low density and good ability to suppress acoustic waves. This application is often present in the automotive industry, where natural fibers are used as a filler in composite elements of vehicle interiors [20]. An important technological limitation in the natural fiber-reinforced composites industry is the temperature, which should not be bigger than 230 °C. Exceeding that temperature would cause the degradation of the fiber. Nevertheless, it is not an issue while using chemo-hardening resins.

The aim of the study was to determine the adhesion force of the polymeric coatings to the steel base, to compare the impact resistance of multilayer coatings based on the damage analysis caused by the impact of the energy of 17 J, to determine the coating resistance to cracking and peeling from the base as well as to investigate the influence of the coating on static mechanical properties and the fatigue lifetime of the sample, which is expected to improve. In general, the most widely used strategy of fatigue lifetime improvement is strengthening metallic structures using CFRP (carbon fiber reinforced polymer) patches [21–23]. The main reason for this is the redistribution of forces in metallic and composite structures. In this paper, the beneficial effect of polyurea composite on the fatigue performance of AISI 304 steel will be also demonstrated.

2. Materials and Methods

As a base material, austenitic steel AISI 304 in the form of 0.5 mm thick metal sheet was used. Chemical composition and static tensile results [24] of AISI 304 ((0.04%C, 1.1%Mn, 0.41%Si, 0.0437%P, 0.0044%S, 18.16%Cr, 8%Ni, 0.0335%Mo, 0.1%V, 0.32%Cu) steel are included in Table 1.

Table 1. Static mechanical properties of the analyzed steel AISI304, based on [24].

Material	Ultimate Tensile Strength UTS (MPa)	Yield Strength $R_{pl}/R_{0.2}$ (MPa)	Young Modulus E (GPa)	Poisson Ratio ν (-)	Vickers Hardness HV (-)	Elongation at Break A_5 (%)
AISI 304 steel	612	312	187	0.29	252	57

The base was degreased with acetone. For the adhesion and impact tests 100 × 100 mm samples were prepared; for static tensile and fatigue tests oar-shaped samples were prepared (Figure 1).

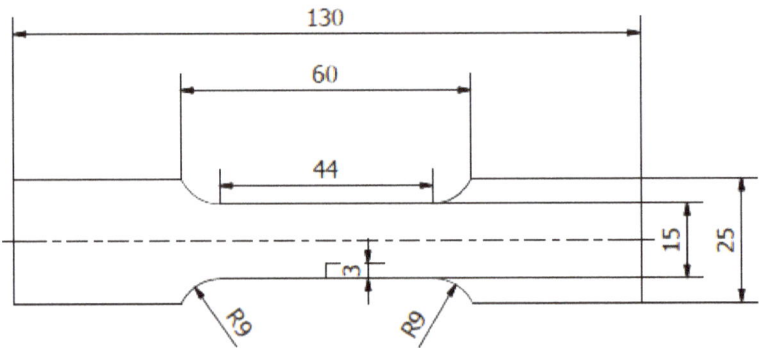

Figure 1. The geometry of a notched sample.

The type of applied coatings and layers configurations are presented in Table 2.

Table 2. Types of applied coatings.

Sample Designation	Composition and Configuration of the Coating
EP	epoxy resin
EP_GF	epoxy resin + glass fabric
EP_GF_HF	epoxy resin + glass fabric hemp fiber
EP_PUA	epoxy resin + polyurea resin
EP_GF_PUA	epoxy resin + glass fabric + polyurea resin
EP_GF_HF_PUA	epoxy resin + glass fabric + hemp fiber + polyurea resin

The coatings were applied manually. All samples were coated with epoxy resin LH 289 and characterized by low viscosity (Havel composites); more information about the resin are presented in Table 3. The first layer of the coating was modified by reinforcing it with glass fiber Areoglas 163 g/m^2 (Table 4) and/or with cut hemp fibers of 40 mm length. Part of the samples was covered with two-component polyurea coating Almacoat Floor Sl (Table 5). Obtained surfaces of the samples are presented in Figure 2.

Table 3. Properties of used epoxy resin.

Molecular weight (g/mol)	180–193
Color	Max.3
Epoxide index, mol/1000	0.51–0.56
Ignition temperature, °C	above 150
Viscosity (mPa, 25 °C)	500–900
Density (g/cm^3)	1.12–1.16

Table 4. Properties of used glass fiber.

Surface mass		160 ± 10 (g/m^2)
Plait		Plain weave
Edges		cut
Matrix density		120 ± 1
Storage	temperature	Up to 25 °C
	humidity	Up to 68%

Table 5. Properties of a used polyurea resin.

Viscosity (25 °C)	ISO-7000 mPas, Polyol-500 mPa	EN ISO 2555 (Brookfield)
Volatiles	0%	-
Density (25 °C)	ISO-1.10 g/cm^3, Polyol-1.05 g/cm^3,	EN ISO 1675
Life time after mixing (20 °C)	9 min	-
Treatment time after effusion (20 °C)	20 min	-
Application temperature	+10 °C to 30 °C	-
Mixing proportions ISO:Polyol	100:13 (weight)	-
Recommended thickness	2 mm	-
Tensile strength	13 MPa	EN ISO 527
Elongation	650%	EN ISO 527
Adhesion to the base (steal)	>5 MPa	EN ISO 4624
Adhesion to the base (concrete)	Rapture in concrete	EN 1542
Shore's hardness	80A	EN ISO 868
Water absorption (7 days)	Up to 3.5%	-

Figure 2. Images of the obtained surfaces: (**a**) EP; (**b**) EP_GF; (**c**) EP_GF_HF; (**d**) EP_PUA; (**e**) EP_GF_PUA; (**f**) EP_GF_HF_PUA.

3. Results

3.1. Adhesion of Coatings

Measurements of the adhesion of the coatings were carried out using a pull-off method, according to the standard PN-EN ISO 4624:2016-05 [25] using the PosiTest AT-A device (DeFelsko Corporation, Ogdensburg, NY, USA). During the test, the pull-off force of the stamp from the polymer coating based on the steel base was estimated. Prior to the test, measurement stamps were applied to the coating. After curing the glue, the circular notch around the stamp was cut and, subsequently, the pull-off test was carried out. There were five measurements done per each type of coating. In Figure 3 images of samples with glued measurement stamps are presented. The pull-off test classification according to the standard PN–EN ISO 4624:2016 [25] is presented in Table 6.

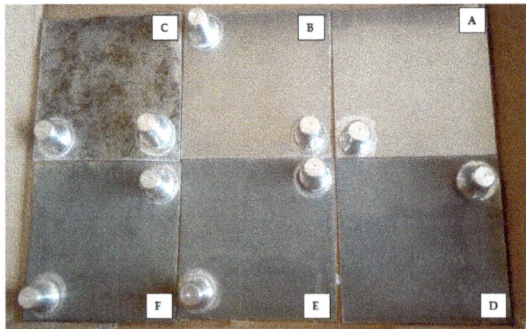

Figure 3. Samples prepared for adhesion testing, A—EP; B—EP_GF; C—EP_GF_HF; D—EP_PUA; E—EP_GF_PUA; F—EP_GF_HF_PUA.

The results of adhesion measurements are presented in Table 7. In all cases adhesive separation between the base and the first layer of the coating was obtained. In Figure 4, the results of the pull-off force measurement obtained during the PosiTest test are shown. Due to the different materials (including fibers) used for the coating, the total thicknesses of the layers were different. However, this did not change the reinforcement and redistribution of stresses, as shown by the results of the static tests. During the application of the layers, it was ensured that the thickness of the layers was equally distributed. Quality control with optical scanners revealed differences in thickness not greater than 8% of the applied layer.

Table 6. Pull-off test classification according to the norm PN–EN ISO 4624:2016 [25].

Designation	Description
A	Cohesive separation in the base
A/B	Adhesive separation between the base and the first layer
B	Cohesive separation in the first layer
B/C	Adhesive separation between the first and the second layer
N	Cohesive separation in the n-th layer of the system
n/m	Adhesive separation between the n-th and the m-th layer of the system
-/Y	Adhesive separation between the last layer and the adhesive
Y	Cohesive separation in the adhesive
Y/z	Adhesive separation between the stamp and the adhesive

Table 7. Results of the adhesion pull-off tests.

Sample Designation	Macroscopic Image	Stresses Occurring between the Stamp and the Sheet (MPa)	Type of Separation
EP		0.22	A/B
EP_GF		0.29	A/B
EP_GF_HF		0.38	A/B
EP_PUA		1.33	A/B
EP_GF_PUA		0.84	A/B
EP_GF_HF_PUA		0.59	A/B

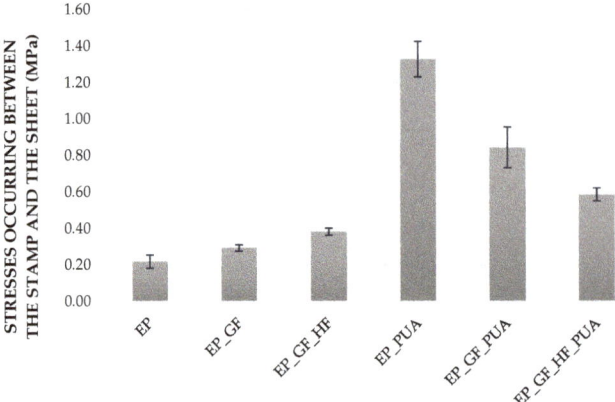

Figure 4. Pull-off force values obtained in the PosiTest test for all types of coating.

3.2. Coatings Impact Resistance

Coatings resistance to cracking or peeling from the base was evaluated according to the standard PN-EN ISO 6272-1:2011 [26] using the impact resistance testing device—TQC (TQC Sheen B.V., Capelle aan den IJssel, Netherlands), presented in Figure 5. The test consists of determining the minimum height of fall for 20 mm diameter mass, under normalized conditions, in order to damage investigated coatings. The research according to this procedure was conducted also in papers [27,28].

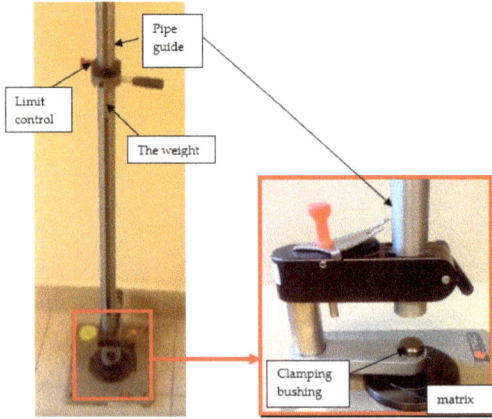

Figure 5. The test stands for evaluation of samples' impact resistance—general view and the magnification of the base of the test stand.

Initial tests were conducted on the metallic base without any coating with a load of 1 kg. The metal sheet was hit from different heights and there was no observed rapture of the material (Figure 6). Due to the lack of visible material damage after dropping the weight of 1 kg from maximal height of 1 m, the weight was changed to 2 kg. Material damage was observed for the drop of 2 kg weight from 0.9 m height. The obtained sample was used as a reference sample for further tests on coated samples (Figure 7).

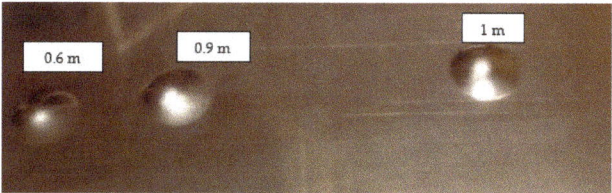

Figure 6. Tested metal sheet deformation after the impact of 1 kg weight from different heights: 0.6, 0.9, and 1.0 m.

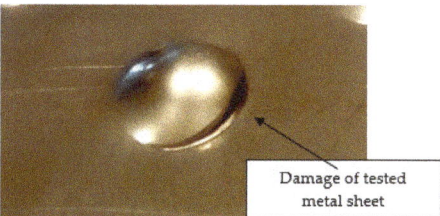

Figure 7. Tested metal sheet damage after the impact of 2 kg weight from 0.9 m height.

The impact resistance test was conducted on previously-prepared coated samples, using the weight of 2 kg gravity dropped from 0.9 m height. Observed results of the test are presented in Tables 8 and 9.

Table 8. Damage observed after the impact resistance test.

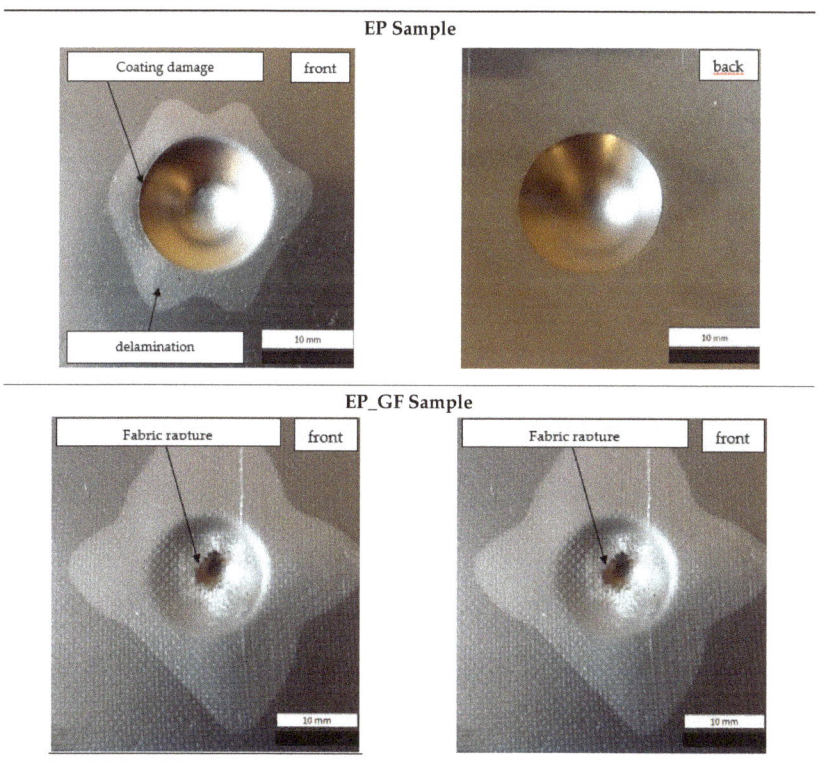

Table 8. *Cont.*

EP_GF_HF Sample	
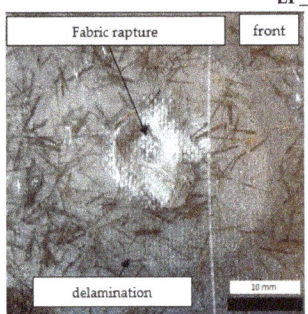 front — Fabric rapture, delamination	 back

EP_PUA Sample	
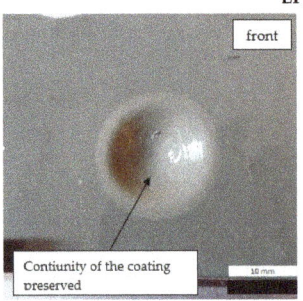 front — Contiunity of the coating preserved	 back

EP_GF_PUA Sample	
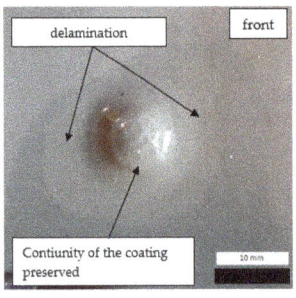 front — delamination, Contiunity of the coating preserved	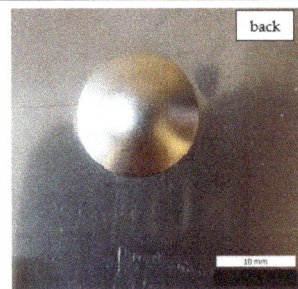 back

EP_GF_HF_PUA Sample	
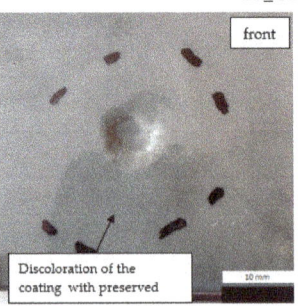 front — Discoloration of the coating with preserved	 back

Table 9. Coating damage after the impact resistance test.

Sample Designation	Delamination Surface Area [mm^2]	Remarks
EP	70	Coating delamination, breakage at the edge of the deformation
EP_GF	152	Coating delamination and glass fiber rapture
EP_GF_HF	196	Coating delamination and glass fiber rapture
EP_PUA	53	Lack of damage and delamination of the coating, continuity preserved
EP_GF_PUA	96	Coating delamination, lack of damage, continuity preserved
EP_GF_HF_PUA	166	Coating delamination, lack of damage, continuity preserved

3.3. Static Tensile Test of Notched Specimens

The static tensile test was conducted on MTS 810 Material Testing Machine (MTS Systems Corporation, Eden Prairie, MN, USA). The test was conducted for samples with EP_GF, EP_PUA, EP_GF_PUA, and EP_GF_HF coatings. The results are the mean of tests on five samples per coating and presented in Figure 8. All results correspond well with the previously obtained [29] experimental data for the same specimen configuration (without coating) made from AISI 304 steel. The critical gross-section tensile stress for the notched (k_t = 5.88) AISI 304 steel specimen was estimated at the level 505 MPa. The results for the sample with EP coating were no different from the value for non-coated steel.

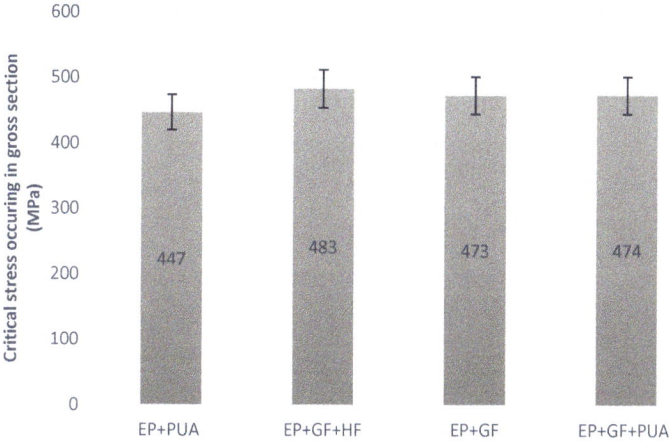

Figure 8. Comparison of the failure load during static tensile test.

3.4. Fatigue Testing

The fatigue test was conducted on the uniaxial MTS 810 Material Testing Machine equipped with a 5 kN load cell under a stress-controlled mode for one selected load level. During the test stress, ratio (R = 0.05) and frequency (f = 20 Hz) were kept constant. All specimens were loaded using sinusoidal waveform with the maximum load level F_{MAX} = 1400 N and minimum load level F_{MIN} = 70 N. In order to achieve proper surface and geometry of the notch, as well as to avoid delamination of the coating, the notch was cut out using the diamond string method. Obtained results are the mean of five samples and presented in Figure 9.

Materials **2019**, *12*, 3137

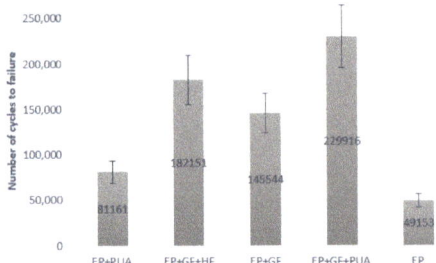

Figure 9. Number of cycles to failure (mean of five) for samples with different coatings.

The macroscopic images of broken specimens are presented in Figure 10.

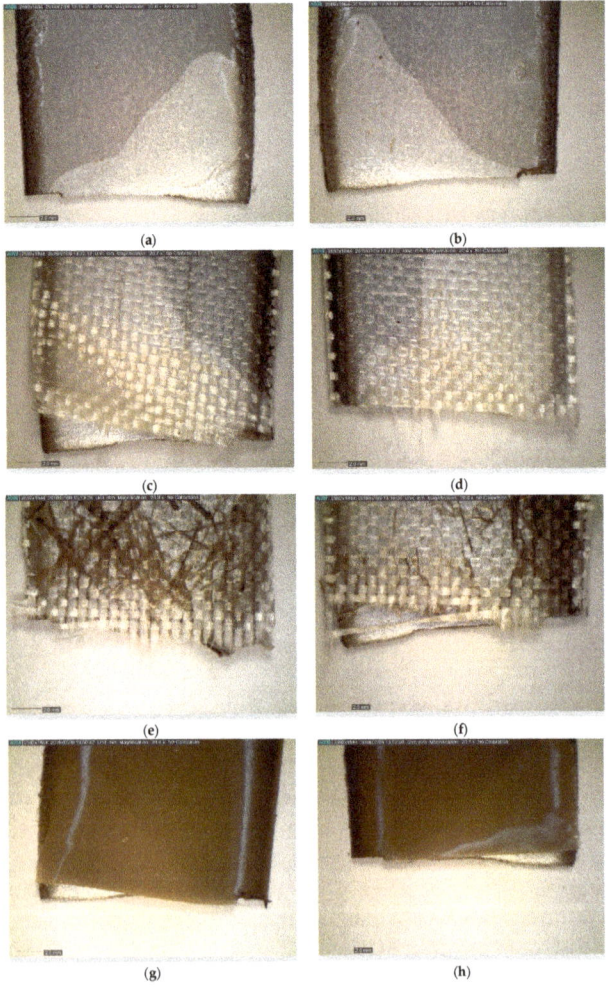

Figure 10. *Cont.*

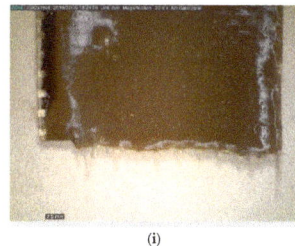
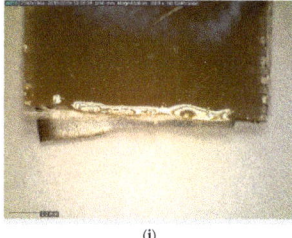

(i)　　　　　　　　　　　　　　　(j)

Figure 10. Images of the fracture area after fatigue test for samples: (**a**,**b**) EP; (**c**,**d**) EP_GF; (**e**,**f**) EP_GF_HF; (**g**,**h**) EP_PUA; (**i**,**j**) EP_GF_PUA.

Coatings with glass fiber as a constituent had a visible asymmetric fracture surface. In case of all samples, there was visible delamination of the coating from the metallic base. The area of delamination was different for each sample, including samples with the same coating. Nevertheless, the delamination process was observed to start behind the notch area, or was even not observed in the notch area at all. This indicates that the notching method was selected and conducted properly. There was no visible delamination between the layers of the coating. The observed break of adhesion forces was, in the case of all samples, between the metallic base and the coating.

4. Conclusions

Based on the performed experimental campaign, the following conclusions can be drawn:

(1) It was observed that the highest value of force required to break adhesive bonds was achieved for the EP_PUA coating, the smallest for the single EP coating.

(2) The largest area of the delamination during the impact test was observed for the EP_GF_HF-coated sample and the smallest for the EP-coated sample.

(3) The static tensile test did not show a significant difference in the influence of the coating on the tensile strength of the material.

(4) Fatigue tests results showed that the difference in the number of cycles to failure depends on the type of coating used. For coatings with polyurea and glass fiber as constituents, the increase of fatigue lifetime was significant.

(5) The macroscopic analysis of the fracture area of damaged samples confirms that the method of notch preparation was correct and had no influence on the behavior of individual samples during the fatigue test.

Due to the possibility of manual application of the coating, if further research on the fatigue lifetime and fatigue crack growth confirm the preliminary results presented in this paper, the coating might be used as an "on-site" fatigue lifetime enhancer and fatigue crack growth retardation tool on the existing structures.

Author Contributions: Conceptualization, J.P. and G.L.; methodology, M.D.; specimen preparation, J.P. and M.D.; data curation, M.D. and J.P.; writing—original draft preparation, J.P. and M.D.; writing—review and editing G.L.; visualization, M.D.; supervision, G.L; project administration, G.L; funding acquisition, G.L.

Funding: The publication has been prepared as a part of the Support Programme of the Partnership between Higher Education and Science and Business Activity Sector financed by City of Wroclaw.

Conflicts of Interest: The authors declare no conflicts of interest.

References

1. Cho, S.H.; White, S.R.; Braun, P.V. Self-healing polymer coatings. *Adv. Mater.* **2009**, *21*, 645–649. [CrossRef]
2. Bertrand, P.; Jonas, A.; Laschewsky, A.; Legras, R. Ultrathin polymer coatings by complexation of polyelectrolytes at interfaces: Suitable materials, structure and properties. *Macromol. Rapid Commun.* **2000**, *21*, 319–348. [CrossRef]

3. Yao, Y.; Liu, N.; McDowell, M.T.; Pasta, M.; Cui, Y. Improving the cycling stability of silicon nanowire anodes with conducting polymer coatings. *Energy Environ. Sci.* **2012**, *5*, 7927–7930. [CrossRef]
4. Laporte, R.J. *Hydrophilic Polymer Coatings for Medical Devices*; CRC Press: New York, NY, USA, 2017.
5. Haag, R.; Wei, Q. Universal polymer coatings and their representative biomedical applications. *Mater. Horizons* **2015**, *2*, 567–577.
6. Ocón, P.; Cristóbal, A.; Herrasti, P.; Fatas, E. Corrosion performance of conducting polymer coatings applied on mild steel. *Corros. Sci.* **2005**, *47*, 649–662. [CrossRef]
7. Sathiyanarayanan, S.; Azim, S.S.; Venkatachari, G. A new corrosion protection coating with polyaniline–TiO_2 composite for steel. *Electrochim. Acta* **2007**, *52*, 2068–2074. [CrossRef]
8. González, M.; Saidman, S. Electrodeposition of polypyrrole on 316L stainless steel for corrosion prevention. *Corros. Sci.* **2011**, *53*, 276–282. [CrossRef]
9. Millet, F.; Auvergne, R.; Caillol, S.; David, G.; Manseri, A.; Pébère, N. Improvement of corrosion protection of steel by incorporation of a new phosphonated fatty acid in a phosphorus-containing polymer coating obtained by UV curing. *Prog. Org. Coat.* **2014**, *77*, 285–291. [CrossRef]
10. Cui, M.; Ren, S.; Qin, S.; Xue, Q.; Zhao, H.; Wang, L. Non-covalent functionalized hexagonal boron nitride nanoplatelets to improve corrosion and wear resistance of epoxy coatings. *RSC Adv.* **2017**, *7*, 44043–44053. [CrossRef]
11. Krasnyy, V.; Maksarov, V.; Olt, J. Improving fretting resistance of heavily loaded friction machine parts using a modified polymer composition. *Agron. Res.* **2016**, *14*, 1023–1033.
12. Madhup, M.; Shah, N.; Wadhwani, P. Investigation of surface morphology, anti-corrosive and abrasion resistance properties of nickel oxide epoxy nanocomposite (NiO-ENC) coating on mild steel substrate. *Prog. Org. Coat.* **2015**, *80*, 1–10. [CrossRef]
13. Lan, P.; Meyer, J.L.; Vaezian, B.; Polycarpou, A.A. Advanced polymeric coatings for tilting pad bearings with application in the oil and gas industry. *Wear* **2016**, *354*, 10–20. [CrossRef]
14. Dong, X.; Jiang, X.; Li, S.; Kong, X.Z. Polyurea Materials and Their Environmental Applications. *IOP Conf. Ser. Mater. Sci. Eng.* **2019**, *484*, 012043. [CrossRef]
15. Toader, G.; Rusen, E.; Teodorescu, M.; Diacon, A.; Stanescu, P.O.; Rotariu, T.; Rotariu, A. Novel polyurea polymers with enhanced mechanical properties. *J. Appl. Polym. Sci.* **2016**, *133*. [CrossRef]
16. Guo, H.; Guo, W.; Amirkhizi, A.V.; Zou, R.; Yuan, K. Experimental investigation and modeling of mechanical behaviors of polyurea over wide ranges of strain rates and temperatures. *Polym. Test.* **2016**, *53*, 234–244. [CrossRef]
17. Pach, J.; Pyka, D.; Jamroziak, K.; Mayer, P. The experimental and numerical analysis of the ballistic resistance of polymer composites. *Compos. Part B Eng.* **2017**, *113*, 24–30. [CrossRef]
18. Myers, J.J. Polyurea Coated and Plane Reinforced Concrete Panel Behavior under Blast Loading: Numerical Simulation to Experimental Results. *Trends Civ. Eng. its Arch.* **2018**, *1*, 001–0012. [CrossRef]
19. Pan, X.; Shi, Z.; Shi, C.; Ling, T.-C.; Li, N. A review on concrete surface treatment Part I: Types and mechanisms. *Constr. Build. Mater.* **2017**, *132*, 578–590. [CrossRef]
20. Pach, J.; Kaczmar, J.W. Influence of the chemical modification of hemp fibers on selected mechanical properties of polypropylene composite materials. *Polimery* **2011**, *56*, 385–389. [CrossRef]
21. Lesiuk, G.; Katkowski, M.; Duda, M.; Królicka, A.; Correia, J.; De Jesus, A.; Rabiega, J. Improvement of the fatigue crack growth resistance in long term operated steel strengthened with CFRP patches. *Procedia Struct. Integr.* **2017**, *5*, 912–919. [CrossRef]
22. Lesiuk, G.; Katkowski, M.; Correia, J.F.D.O.; De Jesus, A.M.; Błażejewski, W. Fatigue crack growth rate in CFRP reinforced constructional old steel. *Int. J. Struct. Integr.* **2018**, *9*, 381–395. [CrossRef]
23. Ghafoori, E.; Motavalli, M. Innovative CFRP-Prestressing System for Strengthening Metallic Structures. *J. Compos. Constr.* **2015**, *19*, 04015006. [CrossRef]
24. Chockalingam, P.; Wee, L.H. Surface roughness and tool wear study on milling of AISI 304 stainless steel using different cooling conditions. *Int. J. Eng. Technol.* **2012**, *2*, 1386–1391.
25. International Standard ISO 4624:2016-05 (E), Paints and varnishes—Pull-off test for adhesion. Third version. 2016.
26. International Standard ISO 6272-1:2011 (E), Paints and varnishes. Rapid-deformation (impact resistance) tests-Part 1: Falling-weight test, large-area indenter. Second version. 2011.

27. Dmitruk, A.; Mayer, P.; Pach, J. Pull-off strength of thermoplastic fiber-reinforced composite coatings. *J. Adhes. Sci. Technol.* **2018**, *32*, 997–1006. [CrossRef]
28. Dmitruk, A.; Mayer, P.; Pach, J. Pull-off strength and abrasion resistance of anti-corrosive polymer and composite coatings. *Int. J. Surf. Sci. Eng.* **2019**, *13*, 50–59. [CrossRef]
29. Zimniak, Z.; Lesiuk, G.; Wiśniewski, W. Supercapacitors electropulsing method for improvement the fatigue resistance of the austenitic steel sheets. *Weld. Technol. Rev.* **2018**, *90*, 859.

© 2019 by the authors. Licensee MDPI, Basel, Switzerland. This article is an open access article distributed under the terms and conditions of the Creative Commons Attribution (CC BY) license (http://creativecommons.org/licenses/by/4.0/).

Article

Computed Tomography-Based Characterization of the Fatigue Behavior and Damage Development of Extruded Profiles Made from Recycled AW6060 Aluminum Chips

Alexander Koch *, Philipp Wittke and Frank Walther

Department of Materials Test Engineering (WPT), TU Dortmund University, Baroper Str. 303, D-44227 Dortmund, Germany
* Correspondence: alexander3.koch@tu-dortmund.de; Tel.: +49-231-755-8424

Received: 28 June 2019; Accepted: 22 July 2019; Published: 25 July 2019

Abstract: The possibility of producing profiles directly by hot extrusion of aluminum chips, normally considered as scrap, is a promising alternative to the energy-intensive remelting process. It has to be taken into account that the mechanical properties depend on the quality of the weld seams between the chips, which arise during the extrusion process. To estimate the influence of the weld seams, quasistatic and cyclic investigations were performed on chip-based profiles and finally compared with cast-based extruded profiles. In order to gain comprehensive information about the fatigue progress, different measurement techniques like alternating current potential drop (ACPD)-technique, hysteresis measurements, and temperature measurements were used during the fatigue tests. The weld seams and voids were investigated using computed tomography and metallographic techniques. Results show that quasistatic properties of chip-based specimens are only reduced by about 5%, whereas the lifetime is reduced by about a decade. The development of the fatigue cracks, which propagate between the chip boundaries, was characterized by an intermittent testing strategy, where an initiation of two separate cracks was observed.

Keywords: hot extrusion; fatigue development; aluminum chip solid state recycling; intermittent computed tomography; alternating current potential drop (ACPD)

1. Introduction

Because of the increasing scarcity of resources, demands with regard to lightweight construction have significantly increased in recent years [1]. In this context, aluminum is particularly suitable because of the excellent strength-to-weight ratio and is becoming more and more popular in lightweight-relevant fields such as the automotive and aerospace industries [2]. A disadvantage is the energy-intensive production of primary aluminum compared to other construction metals [3,4]. With a requirement of about 200 GJ per ton, the production of primary aluminum is one of the most energy-intensive production processes [3] and thus exceeds steel production by a factor of ten [4]. To reduce this large amount of energy, more use is being made of secondary aluminum. Conventionally, the aluminum is melted for recycling [5]. A promising alternative with significantly lower energy consumption is solid state recycling by hot extrusion, in which aluminum scrap can be formed directly into profiles. Compared to the remelting process, direct recycling enables a reduction of energy up to 31% [6]. The use of such scrap like chips additionally has the distinct advantage of a reduced price compared to raw aluminum as well as a less material loss due to the high demand of oxides on the surface of the chips [6]. Stern developed the procedure in 1944 [7], albeit the process has only been intensively studied in the last two decades [8–10].

Regarding the mechanical properties of chip-based profiles, it is questionable to what extent the profiles produced by this way meet the requirements in terms of strength and durability. As previous studies show [11,12], the mechanical properties of chip-based profiles depend on the quality of the weld seams occurring between the chips during the extrusion process. The bond strength of the welded aluminum chips mainly depends on time of the contact of the chips and the temperature, as diffusion was found out to play a significant role in the bonding mechanism [12]. On the one hand, the oxide layers have to be broken up in order to enable a direct metal-to-metal contact. On the other hand, the distance where the surfaces are in contact has to be long enough in order to transfer enough energy for the diffusion process, which will be strongly activated by high pressure and temperature. As Cooper and Allwood [13] showed, the influence of the temperature cannot only be explained by the influence on the flow stress of the aluminum, but also the high dependence of the diffusion on the temperature. Because of this, below room temperature process time does not have a significant influence on the bond strength, as diffusion is not prevalent for such temperatures [13].

In the case of a hot extrusion process parameters such as shear stress, pressure, and local strain during the extrusion are critical for a sufficient break-up of the oxide layers and therefore a satisfactory diffusion and welding process [11,12,14]. These parameters can be adjusted by process parameters, especially the extrusion ratio, the ram speed, and the die design [11].

Most known methods to achieve the required process parameters are attributed to the SPD (severe plastic deformation) method [15], which realizes the break-up of the oxides via high local strains and high hydrostatic pressures. While hot extrusion process has often been used to realize the high local strain [11,14,16], other methods have also been investigated such as friction stir extrusion processes [17,18], which have undergone a change in microstructure and hot and cold cracking. Approaches using a compression process at room temperature [9,19] did not lead to success because the shearing forces were too low. Instead, an additional forming process, such as a rolling process [20] was necessary. To significantly increase the local strains, ECAP (equal channel angular pressing) processes were also used which significantly increase the ductility of the resulting profiles and cause grain refining [21]. However, a disadvantage is a more complex process control and significantly increased forces.

Previous investigations [11,12] particularly address the quasistatic properties of chip-based profiles. First investigations regarding cyclic properties [22,23] show a crack propagation along the chip boundaries so that these act as weak links. This leads to a reduction of the lifetime up to a factor of ten [22,23]. A summary of studies regarding different methods of solid state recycling is given by [21], whereby [15] summarizes recent results of mechanical investigations as well as influencing parameters.

The aim of this study is therefore to investigate the mechanisms that lead to the reduced lifetime observed in [22]. In this context, quasistatic and cyclic investigations are used to identify the possible parameters influencing the mechanical properties of extruded chip-based profiles. The focus lies on constant amplitude tests, with stress amplitudes, estimated by load increase tests, which are described in [24]. All specimens were analyzed by X-ray computed tomography (CT) before the tests in order to be able to detect possible influences of defects like pores and delaminations on the mechanical properties.

In order to draw conclusions on the damage development under constant as well as variable amplitude load, supportive measurands, such as the plastic strain amplitude, the change in temperature and the change in alternating current (AC) potential were used. In this context, the alternating current potential drop (ACPD)-technology is the most comprehensive, as it depends on the temperature, the geometry of the specimen, the microstructure, and the defect structure [24,25]. Based on the measured material response, time-dependent microstructural processes that lead to fatigue fracture can be followed.

In order to be able to comprehensively analyze the development of damage during cyclic loading, intermittent fatigue tests were carried out. Therefore, fatigue tests were interrupted at certain numbers of cycles with specific material reactions and characterized by computed tomography in order to determine the crack propagation.

2. Experimental Methodology

2.1. Material and Process Route

In order to characterize the quasistatic and cyclic behavior of directly recycled profiles, chips made of EN AW6060 aluminum alloy were used as a basis for the experiments. The chemical composition was determined by Otto Fuchs Dülken GmbH (Viersen, Germany) by X-ray fluorescence analysis (XRF) and is given in Table 1. The values determined are within the specified limits according to standard (DIN EN 573-3).

Table 1. Chemical analysis of AW6060 aluminum cast alloy (wt.%).

Ref.	Si	Fe	Mn	Mg	Zn	Ti	Al
DIN EN 573-3	0.3–0.6	0.1–0.3	<0.1	0.35–0.6	<0.15	<0.1	Bal.
XRF	0.4	0.21	0.04	0.42	0.01	0.01	Bal.

In this study, the usability of chips for extrusion and the effects on the mechanical properties of chip-based specimens were investigated and compared with conventional cast-based material. To this end, the cast-based and chip-based material were hot extruded with the same flat-face die for comparison purposes. The process route for the production of the chip-based profiles consists of four stages and is shown in Figure 1.

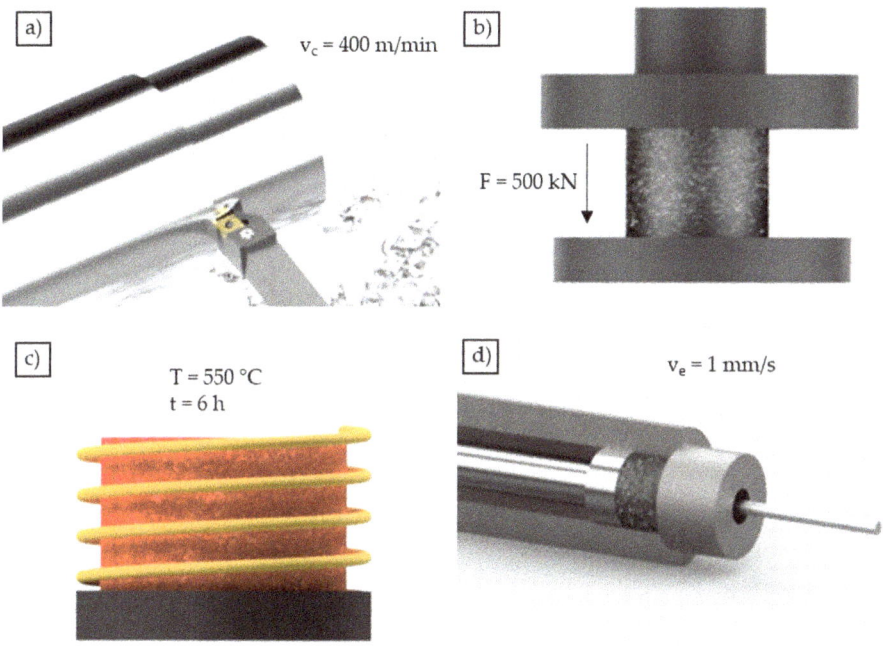

Figure 1. Schematic representation of the extrusion process steps: Machining of the chips (**a**), pre-compaction (**b**), heat treatment (**c**), hot extrusion (**d**).

To produce the extruded profiles, the chips were first produced using AW6060 cast bars by a machining process at the Institute of Machining Technology (ISF) at TU Dortmund University. The geometry of the chips was modified based on relevant research work of Haase et al. [14] and

Güley et al. [11] in such a way that they can be expected to have best welding properties resulting in best mechanical properties. The spirally shaped chips have a length of $l_c = 11.0 \pm 1.7$ mm, a width of $w_c = 7.6 \pm 1.2$ mm, and a thickness of $t_c = 1.1 \pm 0.4$ mm [14]. For the machining of the chips, a cutting insert made by Sandvik (VBMT 160404-UR4225) was used and the chips were produced using a cutting speed of $v_c = 400$ m/min, a feed rate of $f_c = 0.4$ mm, and a cutting depth of $a_p = 2.25$ mm [14].

To ensure enhanced chip surfaces, the chips were cleaned from machining lubricant as well as contaminants and dried after production. In order to ensure satisfactory welding of the chips in the extrusion process, the chips were pre-compacted by a single stroke process at room temperature, using a compaction force of $F = 500$ kN. Finally, the compacted blocks, with a mass of $m_B = 550$ g, a length of $l_B = 92$ mm, a diameter of $d_B = 60$ mm, and a relative density of 78% were homogenized for 6 h at 550 °C, preheated to reduce the necessary extrusion force and then extruded. In order to ensure comparable circumstances, the cast-based material underwent the same heat treatment (homogenization) and the same extrusion parameters as the chip-based billets.

The individual extrusion process parameters were also investigated in detail in [11,14] and optimized with regard to the resulting properties of the extruded profiles. Therefore, the blocks were heated up to a temperature of $T_B = 550$ °C and were extruded with a ram speed of $v_e = 1$ mm/s using a Collin LPA250t hydraulic extrusion press with a maximum ram force of 2.5 MN. The tool was heated up to a temperature of $T_T = 450$ °C. The flat-face die used for the extrusion process had a diameter of $d_d = 12$ mm which results in an extrusion ratio of $R_p = 30.25$. The extrusion ratio is defined as the quotient between the diameter of the billet ($d_B = 66$ mm) and the diameter of the resulting profile ($d_d = 12$ mm). The extruded profiles were cooled by compressed air after leaving the die.

2.2. Metallography

In order to be able to correlate the microstructure with the mechanical properties, a more precise knowledge of the chip orientation and the grain structure, which directly affects the strength according to the Hall-Petch relationship [26], is of importance. Therefore, cast-based and chip-based profiles were cut and cold-embedded perpendicular to the extrusion direction. The profiles were then ground and polished up to a grit size of 0.1 μm using SiO_2 polishing suspension. The microstructure was characterized on cross-sections by means of an electrolytic etching according to Barker [27]. Fluorophosphoric acid (35%) was used as an electrolyte at a flow rate of 12 L/min. On the profile, poled as the anode, a layer applied by the etching process which enables the detection of the grain orientation. The etching was carried out for 90 s at a DC voltage of 20 V using an electrolytic etching device (Struers LectroPol-5, Willich, Germany). The subsequent microstructural characterization under polarized light was carried out on a light microscope (Zeiss Axio Imager M1m, Jena, Germany). Subsequently, the grain size was determined by the linear intercept method.

2.3. Fractography

In order to be able to characterize in particular the deformation and crack propagation behavior of the specimens, the fracture surfaces of the tested specimens were examined in a scanning electron microscope (SEM) (Tescan Mira 3 XMU, Brno, Czech Republic). For a comprehensive characterization of the fracture surfaces, both the information of the element-sensitive backscattered electron detector and the secondary electron detector suitable for topological information were evaluated. Previously, the fracture surfaces were cleaned in an ethanol-filled ultrasonic bath. The investigations were intended to detect stress-dependent changes in the type, shape, and size of cracks and to determine differences in the fatigue behavior between the cast- and chip-based specimens. For the chip-based specimens, knowledge about the preferred crack propagation direction and the role of the welded chips in the fatigue process, as well as their interaction, was gained.

2.4. Mechanical Testing

2.4.1. Tensile Tests

All tensile tests were carried out strain-controlled according to DIN EN ISO 6892-1 at room temperature on a universal testing machine (Instron 3369, High Wycombe, UK) equipped with a load cell with a maximum force of 50 kN. After the machining process, the specimens had a roughness of R_z = 25 µm in the gage length area. Before the tests, the gage length areas of the specimens were ground by means of abrasive paper and then polished using polishing paste.

For strain measurement, a tactile extensometer (Instron type 2630-106) with a gage length of 25 mm and a maximum extension of +50%, −10% was used. The specimen geometry, according to DIN 50125 (tensile test DIN 50125-A 7 × 35) is shown in Figure 2b. The tests were carried out using a strain rate of 0.00025 s^{-1} in the elastic and 0.001 s^{-1} in the elastic-plastic range. The transition to the elastic-plastic range was assumed when exceeding a normal stress of 50 MPa.

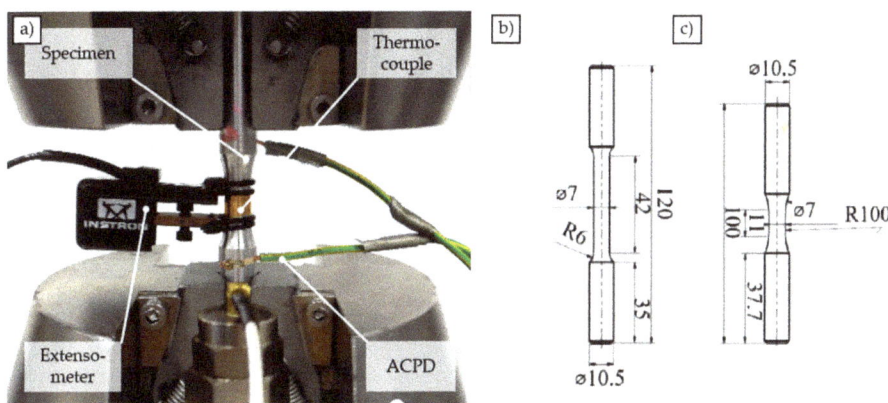

Figure 2. Experimental setup for fatigue experiments (**a**), specimen geometry for quasistatic (**b**), and fatigue (**c**) tests, all units in mm.

2.4.2. Fatigue Tests

Various fatigue tests were carried out in order to determine the load-dependent deformation and damage behavior. Therefore, continuous load increase tests were carried out, described in [24] as a basis to estimate suitable stress amplitudes for the constant amplitude tests. The geometry of the specimens used for this purpose is shown schematically in Figure 2c. All fatigue tests were carried out on a servohydraulic testing machine (Instron 8872) with a load cell with 10 kN load capacity. The tests were performed without superimposed mean stress with a stress ratio of R = −1, a test frequency of f = 10 Hz, and a sinusoidal load–time curve. Specimens, which exceeded N_l = 2 × 10^6 load cycles were defined as run outs.

In order to follow the material reactions, the characteristics of the stress–strain hysteresis were detected by means of a tactile extensometer (Instron type 2620-603, l_0 = 10 mm), the change in electrical resistance by using ACPD (alternating current potential drop) technique, as well as the deformation and damage induced change in temperature by means of thermocouples. In addition to the thermocouple attached to the specimen, the ambient temperature was recorded by three thermocouples placed at different areas in the vicinity of the specimen. As a variable room temperature, which is included in the calculation of the temperature changes of the specimen, the mean value of the temperature measurements of these three additional thermocouples was used. To measure the microstructure sensitive change in AC (alternating current) potential, Matelect CGM-5 system (Harefield, UK) was used. The electrical contacts were spot welded to the specimens, while the poles of the current introduction and the poles of the measurement of the potential were welded each crosswise to reduce interference effects. The current was kept constant at a value of I = 1.7 A, with a signal gain of 90 dB. The frequency f_{AC} was found out to be optimal at a value of f_{AC} = 0.3 kHz. The experimental setup is shown in Figure 2a.

2.5. Computed Tomography-Based Defect Analyses

For the analyses of the internal defect structure as well as the defect development of the cast- and chip-based specimens under cyclic loading, computed tomography (CT) examinations were performed using Nikon XT H 160 X-ray computed tomography scanner (Leuven, Belgium). In order to correlate the defect characteristics as well as the defect distribution with the quasistatic and cyclic properties, all specimens were examined by CT before testing. The volume reconstructions and defect analyses of the CT scans were realized using VGStudio Max 2.2 software. To ensure the comparability of the results of the defect analyses, all specimens were investigated with the same scanning parameters. The parameters which were found to be optimal with regard to the expected image quality are summarized in Table 2.

Table 2. Parameters and settings of computed tomography (CT) examinations for measurements in the gage length area of tensile and fatigue specimens.

Exposure Time	Number of Frames	Beam Intensity	Beam Current	Beam Power	Resolution
250 ms	8	135 kV	98 µA	13.2 W	13.5 µm

In addition to the defect analyses, investigations of the damage development of the extruded chip-based specimens were tracked intermittently. For this purpose, a chip-based specimen was loaded with a certain number of load cycles and then analyzed by CT. Doing so, changes in the internal defect structure, as well as preferred crack initiation sites and propagation direction, were characterized. The fatigue test was each interrupted when significant changes in the material reactions occurred.

3. Results and Discussion

3.1. Metallographic Investigations

After barker etching, the cross-section of the cast-based profile (Figure 3a,b) shows an inhomogeneous grain size distribution in radial direction. In the marginal areas of the cast-based profile significantly smaller grains can be recognized, whereas the grains in the middle of the profile are much larger. Overall, the grains in all areas show a round and uniform shape.

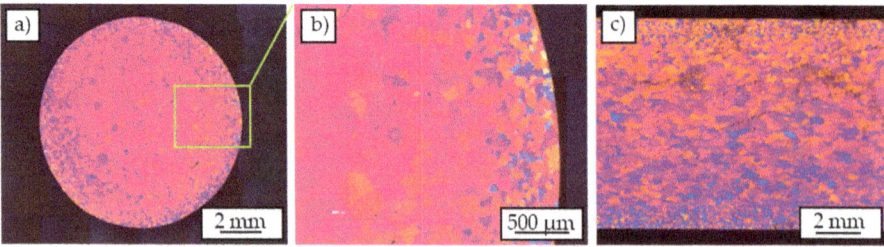

Figure 3. Micrographs of Barker-etched cast-based profile: overview of cross-section (**a**), detailed view of cross-section (**b**), and overview of longitudinal section (**c**).

The optical micrograph of the longitudinal section (Figure 3c) also shows the described inhomogeneity with respect to the grain size distribution and a round shape. The mean grain size in the center of the profile was determined by means of the linear intercept method to be 380 ± 58 µm and 82 ± 37 µm in the marginal area.

Figure 4 shows the images of the Barker-etched chip-based profile. The individual chips are oriented similar, indicated by the same color in polarized light. Solely in the outer area at a diameter >8 mm very different orientations can be identified. Unlike the corresponding images of the cast-based profile (Figure 3), the grains are much more pronounced and separated from each other by clear, black interfaces, so that it can be determined that these are the interfaces of the welded chips.

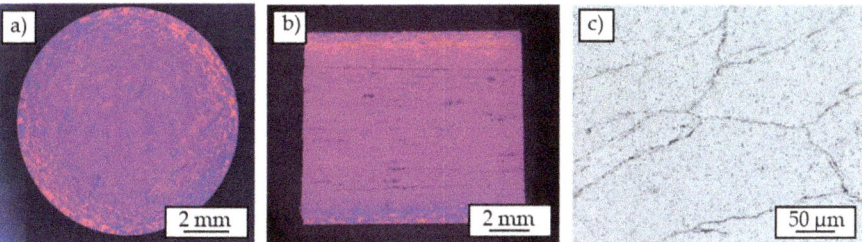

Figure 4. Overview of Barker-etched cross-section (**a**), Barker-etched longitudinal section (**b**), and detailed view of welded chips (**c**).

For the grain structure in the chip-based profile, three different zones can be distinguished (Figure 4a). The first innermost zone is characterized by a different orientation of the individual chips. The grain boundaries correspond to the chip boundaries so that every chip contains a new grain and the grain visibility is high. Since the local strain of the chips in the center of the extrusion ram is low compared to the outer areas, the oxide break-up is insufficient, despite of the high pressure in this area during the extrusion process. This first zone is followed by a second zone in which a large number of very small, differently oriented grains can be detected in the individual chips. In the outermost zone, areas of the same grain orientation also run across the chip boundaries. This can be explained due to the process-related heat input. According to Güley et al. [11], during the extrusion process, areas of huge temperature differences can be identified in extruded profiles. Particularly in the outer areas, high energy inputs with associated temperature increases can be observed due to the frictional conditions prevailing on the die and in the dead metal zone [28]. As a result, these temperature increases lead to local exceedances of the recrystallization temperature and thus to the formation of new grains. Additionally, the local shear stress is high enough to enable recrystallization beyond chip boundaries and therefore sufficient welding of the chips, despite the local pressure decreases to zero in these regions [12]. Because of the subsequent cooling with compressed air, the cooling rates in the outer profile areas are higher. According to Liang et al. [29] higher cooling rates lead to the formation of smaller grains due to the high undercooling achieved.

The elongated grains in the chip-based profile are a result of incomplete recrystallization. Compared to the cast-based material the input of process heat is decreased because the chips are not welded at all. For this reason, the recrystallization temperature in the middle of the profile is not exceeded. In outer areas, the temperature then exceeds the recrystallization temperature, resulting in the formation of sub-grain boundaries within the individual chips. In areas further out, the temperature is then sufficiently high to exceed the recrystallization temperature and thus to cause the formation of new grains, even beyond the chip boundaries.

As already stated, Güley et al. [11] identified two critical parameters influencing the welding process of the chips. The first parameter regards to a critical shear stress above which the encasing oxide layers break down and enable metal-to-metal contact in consequence. As the second parameter, a critical path length is defined, which is understood as a minimum length of the contact of the surface of the chip surfaces in the process which is necessary to allow sufficient welding. Only if both conditions are met sufficient, a successful welding process can be achieved during the extrusion process. At least the parameter of the critical shear stress is directly influenced by the extrusion ratio. Thus, with a larger extrusion ratio, the effective shear stress is increased [14,15]. Because of the friction in the contact area between the profile and the die, the shear stress is significantly higher in the outer regions of the profile than in the central regions. For flat face dies, Haase et al. [14] were able to show that chip delamination phenomenon occurs at a lower extrusion ratio than 17.4. For the chip-based profiles, the extrusion ratio of R_p = 30.25 appears to be sufficient for adequate welding of the chips. Apparently, because of the effective shear stresses in the outer regions of the chip-based profile there is sufficient welding, which is why no delaminations can be found in these regions. Starting from a critical radius, the influence of the friction between the material and the die has then dropped to such an extent that sufficiently high shear stresses can no longer act to break up the oxide layers, resulting in the observed high visibility of the chips.

3.2. Results of Tensile Tests

The results of the tensile tests, summarized in Table 3, clearly show differences between cast- and chip-based specimens. While the tensile tests performed on cast-based specimens show a lower scattering, the chip-based specimens differ more in the results with regard to ultimate tensile strength and yield strength. The cast-based specimens have both a higher ultimate tensile strength and a higher elongation at break than the chip-based specimens. On the other hand, the cast-based specimens show lower values for the yield strength than the chip-based specimens. The lower ultimate tensile strength of the chip-based specimens can be explained by the insufficient quality of the weld seams. In ddition to the reduction of the strength caused by the weld seams, the defects in the specimens in the form of delaminations also reduce the strength due to their notch effect. In general, there are clear differences between the defect sizes in the chip-based specimens which explain the higher scattering of the quasistatic properties.

Table 3. Material characteristics obtained from tensile tests.

Characteristic Value	Cast-Based	Chip-Based
0.2%-yield strength $\sigma_{y,0.2}$ (MPa)	45.9 ± 0.5	54.1 ± 5.4
Ultimate tensile strength σ_{UTS} (MPa)	140.5 ± 1.7	133.3 ± 5.8
Elongation at break ε_f (%)	26.6 ± 2.9	18.2 ± 0.6

The higher 0.2%-yield strength can be explained by the hardening characteristic of the material. Many investigations indicate a pronounced cyclic hardening of AW6060 [30,31]. Thus, when extruding the cast-based material already at the beginning of the extrusion process a material cohesion is given so that no additional deformations of the material are required. Regarding the chip-based material, on the other hand, material cohesion has to be created by local forming of the chips. In this way, the chip-based material has already experienced a significantly higher deformation and thus strain hardening compared to the cast-based material after the extrusion.

In order to investigate the damage mechanisms in case of tensile load, CT analyses of the specimens tested in tensile test were performed. The parameters of the tensile tests were chosen based on the force drop in such a way that the specimens do not fail completely during the test.

The volume reconstruction of a cast-based specimen tested in the tensile test (Figure 5) shows a significant constriction of the specimen just before the fracture. The diameter of the specimen has been reduced in this range from initially d = 7 mm to a value of d = 2.5 mm.

The volume reconstruction of a chip-based specimen (Figure 6) also shows a significant constriction. The diameter is reduced to a value of d = 2.7 mm. Furthermore, cracks are already visible. These propagate between individual chips and effect a separation of the material in the plane of the smallest cross-section. The cracks are rather short and located in the lower left area of the cross-section.

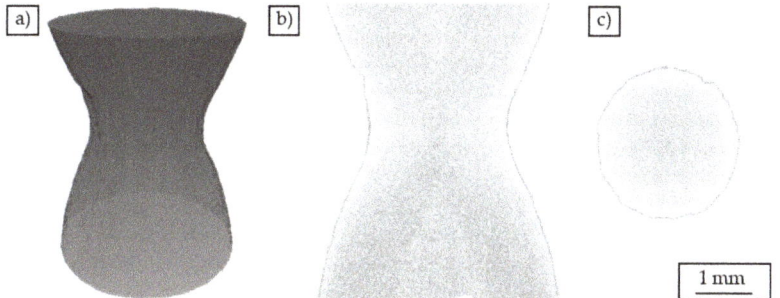

Figure 5. Volume reconstruction of cast-based specimen tested in a tensile test: (**a**) three-dimensional representation, (**b**) cross-sectional view parallel to the load direction, (**c**) cross-sectional view perpendicular to the load direction.

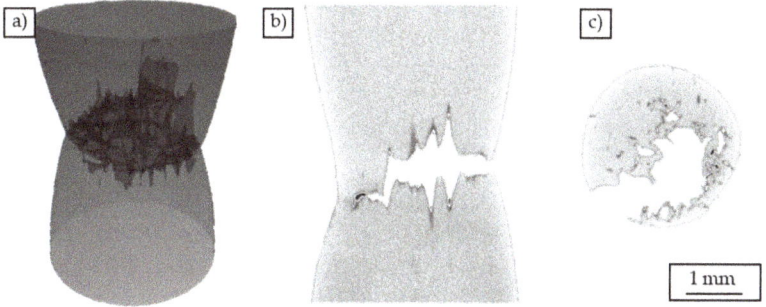

Figure 6. Volume reconstruction of chip-based specimen tested in a tensile test: (**a**) three-dimensional representation, (**b**) cross-sectional view parallel to the load direction, (**c**) cross-sectional view perpendicular to the load direction.

3.3. Results of Fatigue Tests

Figure 7 shows the results of the load increase test (LIT) for the cast-based specimen as well as the chip-based specimen.

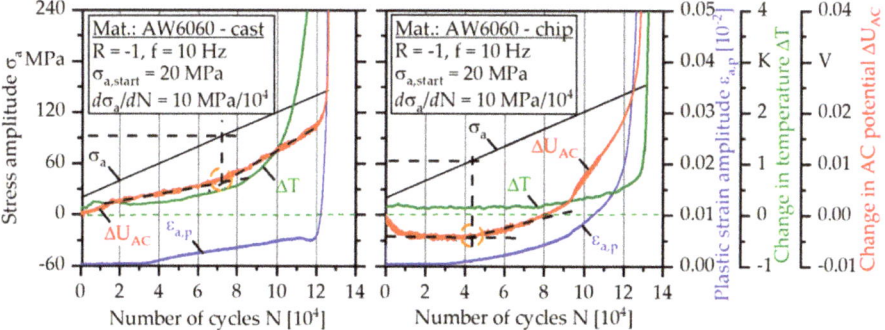

Figure 7. Load increase tests of cast-based and chip-based specimens.

Based on the material response caused by the continuously increasing stress amplitude, the fatigue strength, as stated in [24], can be well estimated. For this purpose, the analysis of the change in

AC potential fits best, as it is the most comprehensive measurement technique. The AC potential is influenced by the temperature, the geometry of the specimen (e.g., changed because of cyclic creep), and especially the microstructure and therefore takes fatigue relevant mechanisms like dislocation accumulation and crack propagation into account [24]. As a result, the course of AC potential follows the courses of the temperature as well as the plastic strain amplitude and additionally takes microstructural changes into account, which cannot be indicated by the temperature or the plastic strain amplitude.

For the cast-based specimen two different regions of linear increase, after a short phase of initial rise of the AC potential, can be distinguished, whereby the slope changes at a stress amplitude of σ_a = 93 MPa. Based on the material response, the fatigue strength can be estimated at about $\sigma_{a,e}$ = 93 MPa. It can be assumed that, above $\sigma_{a,e}$, damage-relevant processes occur in the material, which effect the changes in the material response. The results fit well to the S-N-curve (Figure 10b), where a run out occurred in a constant amplitude test (CAT) at a stress amplitude of σ_a = 90 MPa.

For the chip-based specimen an initial decrease of ΔU_{AC} can be observed until σ_a = 35 MPa is reached due to a compaction of the weld seams presumably. In the stress amplitude region between σ_a = 35 and 63 MPa, the change in AC potential shows a plateau phase, followed by an exponential increase. Analogously to the cast-based specimen the fatigue strength can be estimated at the end of the first linear region at a stress amplitude of about $\sigma_{a,e}$ = 63 MPa. At a stress amplitude of σ_a = 115 MPa, a change in the slope can be observed in the course of the plastic strain amplitude as well as in the change of the AC potential. This is due to the initiation of a second main crack on the opposite side of the first crack, which can be observed by intermittent fatigue tests for most of the chip-based specimens (Figure 9). The yield strength fits well to the stress amplitude of the first increases of the plastic strain amplitude (Table 2). The drastic increase of all measurands for the cast-based and the chip-based specimen at the end of the tests indicates the final crack propagation stage.

Based on the results of the LIT, suitable stress amplitudes for constant amplitude tests for the chip-based material were chosen. As the fatigue strength was estimated to be about 63 MPa, for reaching the high cycle fatigue (HCF)-region a stress amplitude of σ_a = 80 MPa was chosen. For reaching the low cycle fatigue (LCF)-region, a stress amplitude near the stress amplitude at break for the LIT was chosen (σ_a = 120 MPa).

In the constant amplitude test with a stress amplitude of σ_a = 120 MPa, a significant cyclic hardening phenomenon, accompanied by a drop in the plastic strain amplitude as well as in the temperature can be recognized for the cast-based specimen (Figure 8) after an initial strong softening in the first ten cycles. In the beginning, the drop in the plastic strain amplitude runs exponentially and after about N = 5000, becomes linear. The total mean strain increases very rapidly in the first N = 50 load cycles due to a different cyclic hardening behavior in tension and compression direction.

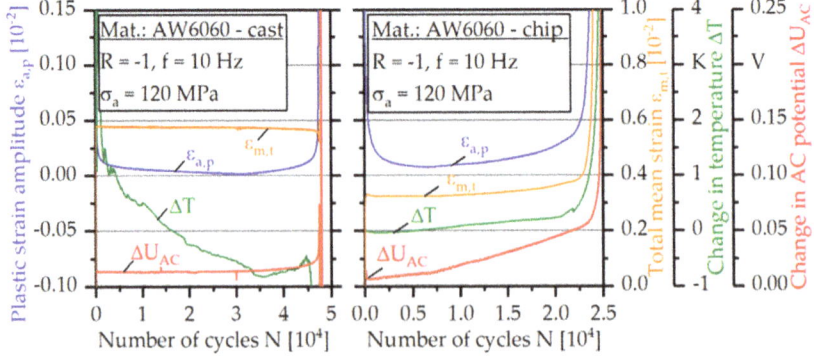

Figure 8. Constant amplitude tests of cast-based and chip-based specimens (σ_a = 120 MPa).

With a number of cycles of about N = 32,000, the material starts cyclic softening, becoming exponentially until break. At the same time, an increase in the temperature is observed. The change in AC potential increases abruptly at the beginning of the test by about ΔU_{AC} = 0.015 V. Subsequently, this rises to a number of cycles of about N = 40,000 initially linear and then exponentially up to the number of cycles to failure N_f = 47,487.

The constant amplitude test of the chip-based specimen tested at the identical stress amplitude of σ_a = 120 MPa (Figure 8) shows a comparable qualitative course for the three measured measurement techniques considered. The failure occurred at a number of cycles of N_f = 24,180. At the beginning of the test, a strong cyclic hardening occurs up to a number of cycles of about N = 7000 detectable by a decrease of the plastic strain amplitude. Associated with this, there is a slight drop in the temperature within the first 1500 load cycles. The change in AC potential shows an even shorter reaction to the cyclic hardening with decreasing decay within the first 100 load cycles.

Subsequent to the descending course of the measured quantities, a cyclic softening of the material occurs, which is accompanied by a linear increase of all the measured quantities considered, until these change into an exponential rise until the break. The temperature measurement reacts at the earliest (from about N = 20,000) with a transition to the exponential increase, while the change in AC potential increases exponentially at the latest (from about N = 24,000).

The total mean strain clearly increases in the first N = 50 load cycles to a value of $\sigma_{m,t}$ = 0.32%. After a phase of approximately constant course, this starts to increase linearly from a number of cycles of about N = 8000 linearly to a value of about $\sigma_{m,t}$ = 0.37% until transition into an exponential increase at a number of cycles of about N = 22,000.

An intermittent CAT was performed at a stress amplitude of σ_a = 110 MPa on a chip-based specimen. CT investigations of the crack progress were performed after a certain number of load cycles. The volume reconstructions (top view) and the corresponding cross-sectional images are shown in Figure 9. In the initial state, a tubular defect due to a delamination between the chips is evident. This is because of insufficient break-up of the oxide layers due to low local strain in the innermost regions of the profile. Thereby, in micrographs chip boundaries are visible (Figure 4). The tubular defect is located in the clamping areas and in the conically extending transition area toward the gage length area. Because of the geometry of the specimen, the gage length area is located within this tubular defect, so that the test area is defect-free except of small, isolated delaminations between the chips. After N = 5000 load cycles, crack initiation and propagation can be recognized in the upper region of the specimen. Crack initiation site is the area where the tubular defect cuts the surface because of the conical shape of the specimen. After N = 11,000 load cycles, the described crack grows to a projected length of about 1.5 mm in the cross-section. Further crack initiation can already be recognized on the opposite side. With an applied number of cycles of N = 17,000 this second crack grows to a projected length of about 6 mm. In turn, the tubular defect acts as the initiation site. After N = 19,500 load cycles both cracks merge into a crack parallel to the loading direction along the chip boundaries.

The course of the change in AC potential (constant amplitude test, Figure 8) correlates well to the results of the crack propagation behavior in the intermittent fatigue test (Figure 9). From the point of discontinuity and the subsequent change of the slope, it is believed that the second crack of the chip-based specimen occurs on the opposite side of the first crack, which leads to an overlap of the crack growth rates and therefore to an increase in the slope. As can be ascertained in the intermittent experiments, crack propagation occurs already after N = 5000 load cycles. Accordingly, crack propagation already occurs at the beginning of the test, since the crack initiation phase occurring in the cast-based specimen is eliminated.

The occurring damage is correspondingly expressed in the progressive course of the change in AC potential. However, crack propagation along the grain boundaries can also be recognized for the cast-based specimens on the basis of the fractographic images (Figure 11), although the grain boundaries do not act as crack initiation sites.

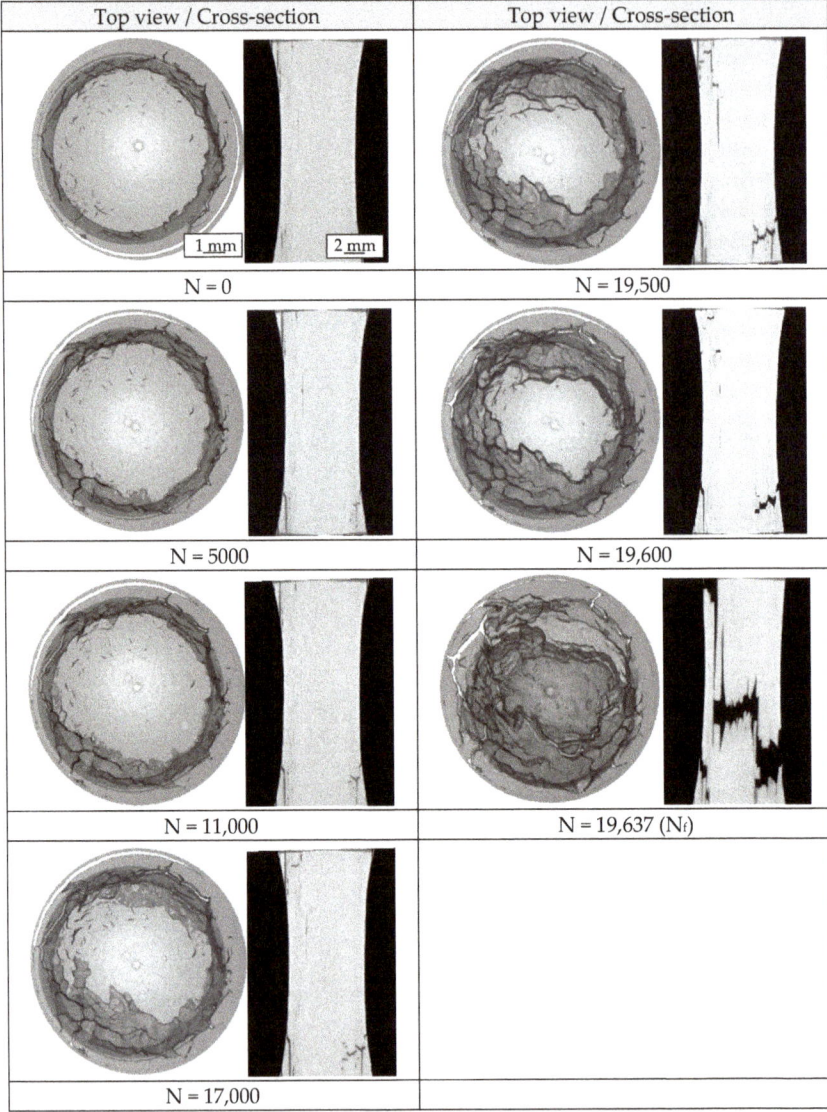

Figure 9. Damage development of a chip-based specimen tested in a constant amplitude test (σ_a = 110 MPa).

In order to compare the change in AC potential, the curves of the cast- and chip-based specimens in CAT at σ_a = 120 MPa are shown in Figure 10a. For better comparability, both axes are normalized. While the cast-based specimen shows a constant course over a longer period of time, for the chip-based specimen a linear increase from the beginning can be recognized. The slope of this linear curve increases from about 0.3 N_f. This can be explained with the initiation of the second crack so that both crack propagation rates summarize. The increase of U_{AC} cannot be explained by the temperature change (Figure 8), since the temperature change is not significant and also does not show a second linear increase with a change of the slope. Analogously, the increase can also not be explained by the total mean strain (Figure 8). Thus it can be clearly seen that an increase of the total mean strain by

$\Delta\sigma_{m,t} = 0.32\%$ causes a change in U_{AC} of only $\Delta U_{AC} = 0.01$ V. Accordingly, the subsequent increasing total mean strain by $\Delta\sigma_{m,t} = 0.05\%$ cannot explain the subsequent large change of the AC potential of $\Delta U_{AC} = 0.04$ V until the beginning of the exponential increase. Since the two influencing variables of the geometry change and the temperature change can be excluded in this way, the only reason for the change in AC potential is the microstructure, especially crack propagation.

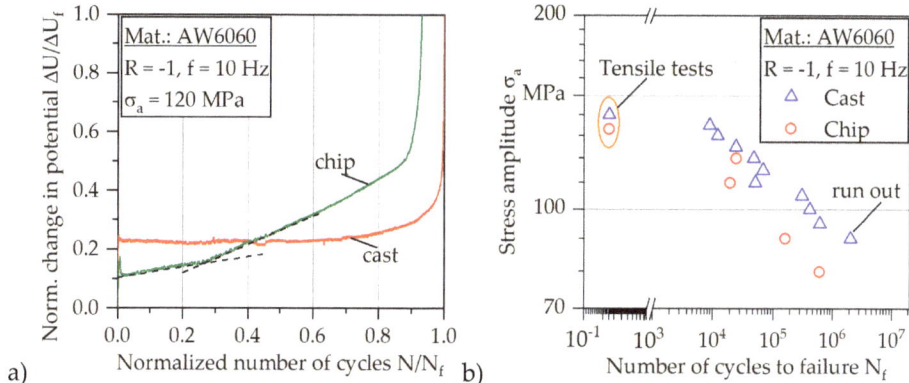

Figure 10. Change in potential for cast-based and chip-based specimens (**a**), S-N curve for cast-based and chip-based specimens (**b**).

On the other hand, it remains unclear why the change in AC potential over a long period of time is linear, since, according to the Paris law [32], an increasing crack rate and thus a progressive increase should be expected. One possible explanation is given by the fiber bridging model [33]. The basis is a barrier effect of individual chip boundaries. These can act as barriers to crack propagation, so that crack propagation is prevented due to the chip boundaries. An identical phenomenon of the linear increase for fiber-based material was found in [33]. Therefore, crack propagation cannot be described using the Paris law.

Fatigue progression begins very early, as crack propagation in the intermittent fatigue test (Figure 9) shows, allowing crack growth after only N = 5000 load cycles. As already described, because of the seam welds present in the chip-based specimens, a crack initiation phase is eliminated, so that it comes directly to the crack propagation phase. In this case, the crack is apparently deflected along the individual chips. After a certain number of cycles, a second crack occurs on the opposite side due to the increase of the stress by the reduced residual area.

Based on the S-N curve (Figure 10), clear stress-dependent differences in the lifetimes of both types of specimens can be recognized. The chip-based specimens show significantly reduced lifetimes compared to the cast-based specimens. While the difference in the HCF-region is about a decade, the differences in the LCF-region are significantly lower. The outlier at $\sigma_a = 110$ MPa for the cast-based specimens is a result of porosity, which can be clarified by means of computed tomography. Overall, the scattering in the cast-based specimens is low. However, lifetime scattering for the chip-based specimens can be correlated with the observed variations in defect sizes.

3.4. Results of Fractography

The SEM-images of the cast-based specimens failed in the fatigue test ($\sigma_a = 120$ MPa) (Figure 11) show two characteristic areas of fatigue fracture and overload fracture which is typical for cyclically tested specimens [34,35]. The fatigue fracture area has a smooth surface covered by striations (Figure 11b). The striations increase in size toward the area of overload fracture. The fatigue crack seems to propagate along the individual grains and thus inter-crystalline. Distinct cracks can be seen between the individual grains, while the surfaces of the grains partly show striations (Figure 11c).

The overload fracture shows a strong ductile deformation. Overall, the fracture mechanism resembles the cup-cone fracture mechanism described in the literature for tensile testing [35]. Preferred crack locations or crack initiation sites cannot be identified.

Figure 11. Scanning electron micrographs of a cast-based specimen tested in a constant amplitude test (σ_a = 120 MPa): overview (**a**), crack course along grain boundaries (**b**), striations (**c**).

Figure 12 shows the SEM-images of a chip-based specimen (σ_a = 120 MPa). Compared to the cast-based specimen (Figure 11) a completely different failure mechanism can be recognized. Analogously to the cast-based specimen, two areas of different shape are apparent. There is the overload area with local separation of the individual chips (Figure 12a). This is particularly pronounced in the central regions of the material, while the outer edge regions appear crack-free in large parts. Visible in this context is the significantly reduced chip width in the peripheral areas.

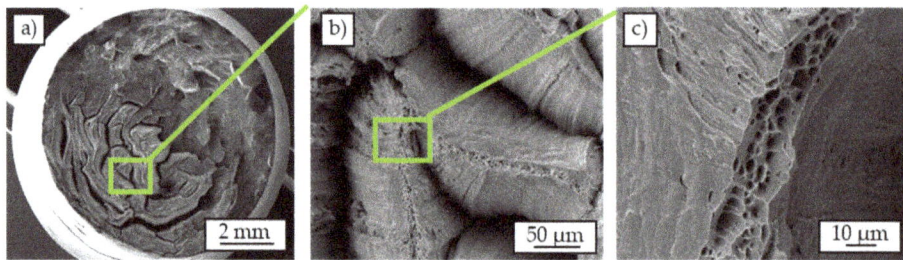

Figure 12. Scanning electron micrographs of a chip-based specimen tested in a constant amplitude test (σ_a = 120 MPa): overview (**a**), detached chips (**b**), fracture pattern on chips (**c**).

In the fatigue fracture area, the chips seem hardly detached compared to the overload fracture area. Similar to the cast-based specimen (Figure 11), individual cracks propagate along the grain boundaries. Considering the individual chips with higher magnification in the first area (Figure 12c), local areas with a ductile honeycomb fracture can be recognized on almost all chip surfaces.

4. Conclusions and Outlook

Within the scope of this work, investigations were carried out with the aim of the mechanical characterization of directly recycled, hot extruded chip-based profiles. The following may be concluded:

(1) The microstructure of chip-based profiles is characterized by three different areas, which originate due to different recrystallization zones.
(2) While micrographs of chip-based profile did not show a delamination between the single chips, it was found that there is a critical diameter where the combination of the necessary properties of high strain and pressure is considered too low to lead to a sufficient welding, which leads to crack initiation, as shown by intermittent test strategy. Therefore, the initiation of the two separate

cracks was observed. Because of the insufficient welding of the chips, the cracks propagate between the chip boundaries.

(3) The weld seams occurring between the chips have a significant influence on the mechanical properties of the resulting profiles. While the quasistatic properties are only slightly reduced by about 5%, the insufficient welded chips in the innermost area of the profile lead to a reduction of the load dependent fatigue life up to one decade.

(4) The load increase test procedure is well suited in order to estimate the fatigue strength with only one specimen.

(5) As the weld seams act as crack initiators, crack propagation phase begins very early for chip-based specimens.

In further studies, concepts of the finite element method will be used to verify by simulation, to what extent the assumption of stress concentration on the opposite side of the first crack is justified. Additionally, the fiber-bridging model will be applied in order to simulate the effects of the chip boundaries on the crack propagation behavior. The findings will then be used to develop a comprehensive material model for chip-based extrusion profiles.

Author Contributions: All tests were performed by A.K.; the figures were prepared by A.K.; the original draft was written by A.K. and P.W.; F.W. supervised the project and reviewed the manuscript.

Funding: The authors thank the German Research Foundation (DFG) for its financial support within the research project "Analysis and extension of the limits of application in metal forming based recycling of aluminum chips" (WA 1672/16-1).

Acknowledgments: The authors acknowledge the financial support by German Research Foundation (DFG) and TU Dortmund University within the funding programme Open Access Publishing. The authors also thank F. Kolpak and A.E. Tekkaya at the Institute of Forming Technology and Lightweight Components (IUL), TU Dortmund University, for the provision of profiles in the context of an excellent scientific cooperation within this research project.

Conflicts of Interest: The authors declare no conflict of interest.

References

1. Apelian, D.; Shivkumar, S.; Sigworth, G. Fundamental aspects of heat treatment of cast Al-Si-Mg alloys. *AFS Trans.* **1989**, *97*, 727–742.
2. Starke, E.A.; Staley, J.T. Application of modern aluminium alloys to aircraft. *Fundam. Alum. Metall.* **2011**, *24*, 747–783.
3. Schwarz, H.G. Aluminium production and energy. *Encycl. Energy* **2004**, *1*, 81–95.
4. Worrell, E.; Price, L.; Martin, N.; Farla, J.; Schaeffler, R. Energy intensity in the iron and steel industry. *Energy Policy* **1997**, *25*, 727–744. [CrossRef]
5. Soo, V.K.; Peeters, F.R.; Compston, P.; Doolan, M.; Duflou, J.R. Economic and environmental evaluation of aluminium recycling based on a Belgian case study. *Procedia Manuf.* **2019**, *33*, 639–646. [CrossRef]
6. Gronostajski, J.; Matuszak, A. The recycling of metals by plastic deformation: An example of recycling of aluminum and its alloys chips. *J. Mater. Process. Technol.* **1999**, *92*, 35–41. [CrossRef]
7. Stern, M. Direct extrusion applied to light metal scrap. *Iron Age* **1951**, *28*, 71–73.
8. Etherington, C. Conform and the recycling of non-ferrous scrap metals. *Conserv. Recycl.* **1987**, *2*, 19–29. [CrossRef]
9. Fogagnolo, J.B.; Ruiz-Navas, E.M.; Simon, M.A.; Martinez, M.A. Recycling of aluminium alloy and aluminium matrix composite chips by pressing and hot extrusion. *J. Mater. Process. Technol.* **2003**, *143–144*, 792–795. [CrossRef]
10. Galanty, M.; Kazanowski, P.; Kansuwan, P.; Misiolek, W.Z. Consolidation of metal powders during the extrusion process. *J. Mater. Process. Technol.* **2002**, *125*, 491–496. [CrossRef]
11. Güley, V.; Güzel, A.; Jäger, A.; Ben Khalifa, N.; Tekkaya, A.E.; Misiolek, W.Z. Effect of die design on the welding quality during solid state recycling of AA6060 chips by hot extrusion. *Mater. Sci. Eng. A* **2013**, *574*, 163–175. [CrossRef]

12. Donati, L.; Tomesani, L. The prediction of seam welds quality in aluminum extrusion. *J. Mater. Process. Technol.* **2004**, *153*, 366–373. [CrossRef]
13. Cooper, D.R.; Allwood, J.M. Influence of diffusion mechanisms in aluminium solid-state welding processes. *Procedia Eng.* **2014**, *81*, 2147–2152. [CrossRef]
14. Haase, M.; Ben Khalifa, N.; Tekkaya, A.E.; Misiolek, W.Z. Improving mechanical properties of chip-based aluminum extrudates by integrated extrusion and equal channel angular pressing (iECAP). *Mater. Sci. Eng. A* **2012**, *539*, 194–204. [CrossRef]
15. Ab Rahim, S.N.; Lajis, M.A.; Ariffin, S. A Review on Recycling Aluminum Chips by Hot Extrusion Process. *Procedia CIRP* **2015**, *26*, 761–766. [CrossRef]
16. Gronostajski, J.Z.; Kaczmar, J.W.; Marciniak, H.; Matuszak, A. Direct recycling of aluminum chips into extruded products. *J. Mater. Process. Technol.* **1997**, *64*, 149–156. [CrossRef]
17. El Mehtedi, M.; Forcellese, A.; Mancia, T.; Simoncini, M.; Spigarelli, S. A new sustainable direct solid state recycling of AA1090 aluminum alloy chips by means of friction stir back extrusion process. *Procedia CIRP* **2019**, *79*, 638–643. [CrossRef]
18. Tang, W.; Reynolds, A.P. Production of wire via friction extrusion of aluminum alloy machining chips. *J. Mater. Process. Technol.* **2010**, *210*, 2231–2237. [CrossRef]
19. Kuzman, K.; Kacmarcik, I.; Pepelnjak, T.; Plancak, M.; Vilotic, D. Experimental consolidation of aluminum chips by cold compression. *J. Prod. Eng.* **2012**, *15*, 79–82.
20. Chiba, R.; Nakamura, T.; Kuroda, M. Solid-state recycling of aluminium alloy swarf through cold profile extrusion and cold rolling. *J. Mater. Process. Technol.* **2011**, *211*, 1878–1887. [CrossRef]
21. Wan, B.; Chen, W.; Lu, T.; Liu, F.; Jiang, Z.; Mao, M. Review of solid state recycling of aluminum chips. *Resour. Conserv. Recycl.* **2017**, *125*, 37–47. [CrossRef]
22. Koch, A.; Wittke, P.; Walther, F. Characterization of the fatigue and damage behavior of extruded AW6060 aluminum chip profiles. In *Structural Integrity: Mechanical Fatigue of Metals, Experimental and Simulation Perspectives*, 1st ed.; Correia, J.A.F.O., Jesus, A.M.P., Fernandes, A.A., Calcada, R., Eds.; Springer International Publishing: Basel, Switzerland, 2019; pp. 11–19.
23. Koch, A.; Henkel, T.; Walther, F. Characterization of the anisotropy of extruded profiles based on recycled AW6060 aluminum chips. In Proceedings of the 3rd International Conference on Structural Integrity and Durability, Dubrovnik, Croatia, 4–7 June 2019; pp. 1–10.
24. Starke, P.; Walther, F.; Eifler, D. Model-based correlation between change of electrical resistance and change of dislocation density of fatigued-loaded ICE R7 wheel steel specimens. *Mater. Test.* **2018**, *60*, 669–677. [CrossRef]
25. Yan, Y.; Yin, H.; Sun, Q.P.; Huo, Y. Rate dependence of temperature fields and energy dissipations in non-static pseudoelasticity. *Contin. Mech. Thermodyn.* **2012**, *24*, 675–695. [CrossRef]
26. Hall, E.O. The deformation and ageing of mild steel: III Discussion of results. *Proc. Phys. Soc.* **1951**, *64*, 747–753. [CrossRef]
27. Tamadon, A.; Pons, D.J.; Sued, K.; Clucas, D. Development of metallographic etchants for the microstructure evolution of A6082-T6 BFSW welds. *Metals* **2017**, *7*, 423. [CrossRef]
28. Eivani, A.R.; Karimi Taheri, A. The effect of dead metal zone formation on strain and extrusion force during equal channel angular extrusion. *Comput. Mater. Sci.* **2008**, *42*, 14–20. [CrossRef]
29. Liang, G.; Ali, Y.; You, G.; Zhang, M.X. Effect of cooling rate on grain refinement of cast aluminium alloys. *Materialia* **2018**, *3*, 113–121. [CrossRef]
30. Adamczyk-Cieslak, B.; Mizera, J.; Kurzydlowski, K.J. Microstructures in the 6060 aluminium alloy after various severe plastic deformation treatments. *Mater. Charact.* **2011**, *63*, 327–332. [CrossRef]
31. Hockauf, K.; Niendorf, T.; Wagner, S.; Halle, T.; Meyer, L.W. Cyclic behavior and microstructural stability of ultrafine-grained AA6060 under strain-controlled fatigue. *Procedia Eng.* **2010**, *2*, 2199–2208. [CrossRef]
32. Pugno, N.; Ciavarella, M.; Cornetti, P.; Carpinteri, A. A generalized Paris' law for fatigue crack growth. *J. Mech. Phys. Solids* **2006**, *54*, 1333–1349. [CrossRef]
33. Lin, C.T.; Kao, P.W. Effect of fibre bridging on the fatigue crack propagation in carbon fibre reinforced aluminum laminates. *Mater. Sci. Eng.* **1995**, *190*, 65–73. [CrossRef]

34. Branco, R.; Antunes, F.V.; Costa, J.D.; Yang, F.P.; Kuang, Z.B. Determination of the Paris law constants in round bars from beach marks on fracture surfaces. *Eng. Fract. Mech.* **2012**, *96*, 96–106. [CrossRef]
35. Scheider, I.; Brocks, W. Simulation of cup–cone fracture using the cohesive model. *Eng. Fract. Mech.* **2003**, *70*, 1943–1961. [CrossRef]

© 2019 by the authors. Licensee MDPI, Basel, Switzerland. This article is an open access article distributed under the terms and conditions of the Creative Commons Attribution (CC BY) license (http://creativecommons.org/licenses/by/4.0/).

Article

PSO-BP Neural Network-Based Strain Prediction of Wind Turbine Blades

Xin Liu [1], Zheng Liu [2,*], Zhongwei Liang [2], Shun-Peng Zhu [3], José A. F. O. Correia [4,*] and Abílio M. P. De Jesus [4]

1. Department of industrial product design, Guangzhou University, Guangzhou 510006, China; designer_liuxin@163.com
2. School of Mechanical and Electrical Engineering, Guangzhou University, Guangzhou 510006, China; lzwstalin@126.com
3. School of Mechanical and Electrical Engineering, University of Electronic Science and Technology of China, Chengdu 611731, China; zspeng2007@uestc.edu.cn
4. INEGI, Faculty of Engineering, University of Porto, 4200-465 Porto, Portugal; ajesus@fe.up.pt
* Correspondence: liu_best@yeah.net (Z.L.); jacorreia@inegi.up.pt (J.A.F.O.C.); Tel.: +86-020-39366932 (Z.L.)

Received: 30 April 2019; Accepted: 6 June 2019; Published: 12 June 2019

Abstract: The full-scale static testing of wind turbine blades is an effective means to verify the accuracy and rationality of the blade design, and it is an indispensable part in the blade certification process. In the full-scale static experiments, the strain of the wind turbine blade is related to the applied loads, loading positions, stiffness, deflection, and other factors. At present, researches focus on the analysis of blade failure causes, blade load-bearing capacity, and parameter measurement methods in addition to the correlation analysis between the strain and the applied loads primarily. However, they neglect the loading positions and blade displacements. The correlation among the strain and applied loads, loading positions, displacements, etc. is nonlinear; besides that, the number of design variables is numerous, and thus the calculation and prediction of the blade strain are quite complicated and difficult using traditional numerical methods. Moreover, in full-scale static testing, the number of measuring points and strain gauges are limited, so the test data have insufficient significance to the calibration of the blade design. This paper has performed a study on the new strain prediction method by introducing intelligent algorithms. Back propagation neural network (BPNN) improved by Particle Swarm Optimization (PSO) has significant advantages in dealing with non-linear fitting and multi-input parameters. Models based on BPNN improved by PSO (PSO-BPNN) have better robustness and accuracy. Based on the advantages of the neural network in dealing with complex problems, a strain-predictive PSO-BPNN model for full-scale static experiment of a certain wind turbine blade was established. In addition, the strain values for the unmeasured points were predicted. The accuracy of the PSO-BPNN prediction model was verified by comparing with the BPNN model and the simulation test. Both the applicability and usability of strain-predictive neural network models were verified by comparing the prediction results with simulation outcomes. The comparison results show that PSO-BPNN can be utilized to predict the strain of unmeasured points of wind turbine blades during static testing, and this provides more data for characteristic structural parameters calculation.

Keywords: wind turbine blade; full-scale static test; PSO-BP Neural Network; strain prediction

1. Introduction

Both the reliability and stability of wind turbine blades affect the safety of the whole machine directly. In order to check the rationality of blade design and verify the safety of manufacturing, static experiments of prototype blades have been performed as a necessary part of the blade certification process [1]. Through static experiments of wind turbine blades, the verification of the designed loading capacity of the blades can be built, and the information about structural characteristics, strain and deformation under the test load can be obtained [2]. Existing literature reports substantial researches on the structural testing of wind turbine blades. For example, Jensen et al. [3] continuously loaded a 34-m-long blade in the flap-direction until the blade failed, recorded the displacements throughout the loading history by local displacement measurement equipment, and found that the peeling of the skin and the box girder were the main cause of blade instability. Through the full-scale fatigue experiment of a 3 MW wind turbine blade directed by IEC 61400-23, Lee et al. [4] found that delamination failure will happen at the blade root and figured out the causes of the delamination failure and problems of the conventional design approach by simulating the situations experienced by the blade. During the conditional monitoring on the trailing edge in a full-scale fatigue experiment of a 2 MW wind turbine blade, Pan et al. [5] found that the stress concentration will lead to delamination between GFRP and the balsa wood, and then proposed a method to increase the core materials in the trailing edge by computing the local stress distribution and stability factors, based on finite element calculations. Lee and Park [6] carried out static testing on a 48.3-m-long blade which had initial static testing and fatigue testing. They found that the blade collapsed when the applied load surpassed 70% of the target value. In addition, Lee and Park [6] proposed an improved laminate pattern to enhance the residual strength of the wind turbine blade.

The references as above are mainly related to the analysis of the failure modes as well as failure causes of wind turbine blades. Moreover, measurement and calculation methods of structural characteristic parameters have also been intensively studied, including the influence of defects and size effect [7–9]. Based on the method of finite differences and an arbitrary beam bending and moment theory, Choi et al. [10] proposed a tip deflection calculation method based on the measured strains data analysis for wind turbine blades, and in order to verify the proposed method, they conducted static testing on a 100 kW wind turbine blade with FBG sensors embedded in its shear web; the average calculation error of the proposed method was proved to be within 2.25%. Before performing a static experiment with a 100 kW wind turbine blade, Kim et al. [11] installed FBG sensors into the bonding line among the shear web and spar cap to collect the strain data and then they found that the collected strain data can be effectively used to evaluate the deflections of the wind turbine blades. Roczek-Sieradzan et al. [12] preformed a full-scale static test of a certain wind turbine blade under the combined loadings of flag and edgewise directions, with the overall and local deformation information measured and recorded during the experimental process and proved that the measurement results can be effectively used to analyze the structural performance of wind turbine blades. On the basis of the digital image analysis technology, Dou [13] proposed a deformation testing method for the full-scale static experiment for wind turbine blades and completed the data measurements and analysis of the three-dimensional deformation field. Shi et al. [14] studied the impact of time-varying environmental temperature and humidity on the test results on the basis of a fatigue experiment of a 1.5 MW wind turbine blade; this research could provide a basis for the development and structure testing of wind turbine blades. Pan [15] studied the influence of structure nonlinearity on the experimental results of full-scale wind turbine blades static testing, analyzing the relationship among bending moment, strain, stiffness, and deflection, among others, and provided a more accurate stiffness data for numerical loading calculations. Yan et al. [16] tested and recorded the frequency, deflection and strain of a 48-m-long wind turbine blade, compared the test results with the design value and found that the error range satisfied the DNV-GL2015 specifications.

The above-mentioned researches have made significant achievements in the analysis of blade failure causes, blade load-bearing capacity, and the measurement methods of the full-scale blade static testing, and there are also some researchers who have investigated the blade structural characteristics analysis from the physics of failure mechanisms [17–22]. However, studies on the correlation analysis of the strain, the applied loads, the loading positions, and displacements in static testing are scarce. Actually, in the static testing, the relationships of the strain with the applied loads, loading positions, displacements, etc., are nonlinear, and the number of design variables is numerous [15,23], thus the calculation and prediction of blade structural characteristics are very complicated. Moreover, the number of measuring points and strain gauges in full-scale wind turbine blade static testing are usually limited, thus the structural characteristics of unmeasured points cannot be directly obtained, so that the static testing has little significance for the calibration of blade design [15]. Considering this problem, the methods of Particle Swarm Optimization (PSO) and Neural Networks are considered in this paper. Actually, there are a few studies regarding Particle Swarm Optimization (PSO) and Neural Networks applied to the wind turbine blades analysis. Andrew Kusiak [24] proposed that Neural Networks improved by PSO were applied in the adaptive control of a wind turbine. Cynthia [25] introduced PSO and Neural Networks optimization methodology to optimize the wind velocity and attack angle of a horizontal axis wind turbine in order to obtain the maximum power coefficient. Milad Fooladi [26] applied Neural Networks improved by PSO to assess the different factors affecting flicker in wind turbines. So, the Neural Networks improved by PSO used to solve the problems of wind turbine blades is effective and efficient. However, there are fewer researches about PSO-BPNN in the wind turbine blades studies. Wang Lei [27] applied PSO-BPNN to perform a structural analysis approximation of wind turbine blades, and the effectiveness of the approach was demonstrated.

As a result of a literature review and the concerns raised above, this paper aims at presenting a study on new strain prediction methods by introducing intelligent algorithms. As mentioned above, PSO-BPNN has significant advantages in dealing with non-linear fitting and multi-input parameters, and the models constructed by PSO-BPNN have better robustness and accuracy [28]; thus, PSO-BPNN has been introduced to predict the strain of wind turbine blades in this paper and a new strain-predictive PSO-BPNN model for full-scale wind turbine blades static behavior to be established. The new model can be used to predict the strain values of the unmeasured points and provide more strain data for structural characteristic parameters calculation.

The structure of this paper is organized as follows: Section 2 introduces the conditions and test procedures for the full-scale static testing of a wind turbine blade; the basic concepts of Neural Networks as well as PSO-BPNN are introduced in Section 3; in Section 4, the strain-predictive method for the central of pressure side based on PSO-BPNN is studied; Section 5 presents the conclusions of this research.

2. Full-Scale Static Testing of a Certain Wind Turbine Blade

2.1. The Wind Turbine Blade Introduction

In this research, a blade produced with a box beam structure for the main girder was used. The matrix material is alkali-free glass fibers impregnated with epoxy resin, and the reinforcing phase material is impregnated glass fibre. The mass of this blade is 15,982 kg with an inherent natural frequency of 1.41 Hz, and the main information regarding the turbine blade is listed in Table 1. In the structure of the wind turbine blade, 0° fibres are applied to increase flap-wise bending stiffness, while ± 45° fibres are applied to increase torsional stiffness. Figure 1 illustrates the structure of the wind turbine blade.

Table 1. The main information of the turbine blade.

Rated power	3.0 MW	Maximum chord length	4.32 m
Design life	20 years	Maximum twist angle	15.6°
Blade length	66.5 m	Mass	15,982 kg
Matrix material	Alkali-free glass fibers impregnated with epoxy resin	Reinforcing phase material	Impregnated glass fibre
Main girder stucture		Box beam (2-ply beam)	

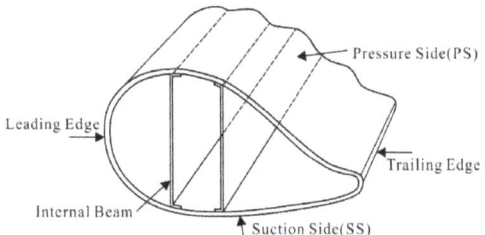

Figure 1. The wind turbine blade structure used in this research.

2.2. Strain Gauges Arrangement

According to Figure 2, 56 strain gauges were attached to the surface of the wind turbine blade along the pressure side (PS), suction side (SS), and the leading edge in addition with the trailing edge, before static testing.

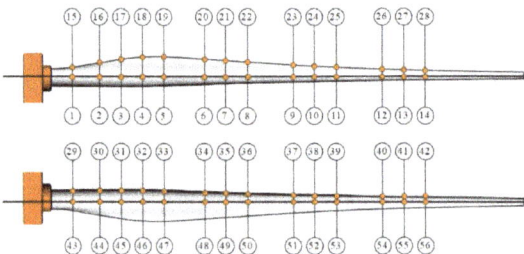

Figure 2. The strain gauges arrangement in the turbine blade.

2.3. Testing Process

In this study, two static tests are successively carried out from different directions, edgewise+ and edgewise−, as shown in Figure 3. This paper will establish the strain prediction models to predict the strain values of the unmeasured points on the center of the pressure side in the two static tests.

Figure 3. Full-scale static experiments of the wind turbine blade. (**a**) Side pulling in the full-scale static experiment of the wind turbine blade; (**b**) Lifting in the full-scale static experiment of the wind turbine blade.

By means of 64 bolts, the root of the blade is installed on the test platform. For the static testing of edgewise+ and edgewise−, five loading positions were chosen and the loading points arranged according the distances of 18.00 m, 30.00 m, 42.00 m, 50.00 m, and 60.00 m measured from the root of the blade, respectively, as shown in Figure 4. It is worth noting that a crane is used to avoid the blade tip landing in the test of edgewise+ for large deflection. In addition, the limit load is performed by the side pull, and the target load of each loading point is shown in Table 2. For every loading point, the loading direction is perpendicular to the normal direction of the loading part. In Figure 4, P1–P5 are the positions of the tensile machine, P6 is the crane, and S1–S5 are the load application points.

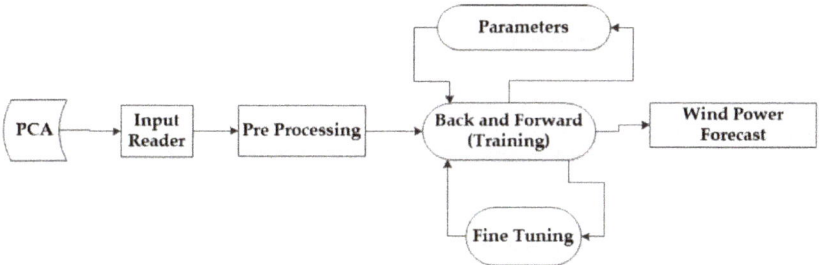

Figure 4. The applied loading diagram of the full-scale static testing.

Table 2. The target load for the five loading positions.

	Distances from the Loading Positions to the Blade Root (m)				
	18.0	30.0	42.0	50.0	60.0
The target load of edgewise+ (kN)	111.5	62.8	45.0	34.5	22.0
The target load of edgewise− (kN)	45.9	81.0	30.0	23.5	24.0

Before starting the test, the applied load, displacement and strain of the blade have been cleared. Then, using the lateral loading device, the blade was loaded by 0%, 40%, 60%, 80%, and 100% of the target load, step by step, while the displacement data and the strain gauge data were recorded during the loading process. After the loading process was completed, the blade was unloaded to the zero state, step by step. The load of each stage on directions of edgewise+ and edgewise− are shown in Tables 3 and 4.

Table 3. The applied load of each stage when loaded on edgewise+.

	The Load Applications (kN)				
Distances from the Loading Positions to the Blade Root (m)	0	40%	60%	80%	100%
18.00	0	44.92	66.81	89.35	111.95
30.00	0	25.64	37.73	50.68	62.88
42.00	0	18.56	27.87	36.07	45.37
50.00	0	14.05	20.88	27.65	34.53
60.00	0	8.88	13.41	17.74	22.11
66.50	81.14	73.87	70.98	68.20	65.35

Table 4. The applied load of each stage when loaded on edgewise−.

	The Load Applications (kN)				
Distances from the Loading Position to the Blade Root (m)	0	40%	60%	80%	100%
18.00	0	19.01	27.60	36.79	46.02
30.00	0	32.69	48.94	65.22	81.65
42.00	0	12.28	18.67	24.29	30.49
50.00	0	9.64	14.18	18.97	23.59
60.00	0	9.62	14.50	19.24	23.99

3. BPNN Improved by PSO

3.1. The Brief Principles of BPNN

The feed-forward BPNN with three layers is proposed for predicting the wind turbine blade static test strain. The first layer is the input layer describing the input data such as values in terms of the distances of points measured from the root of the blade, the applied load of each stage in flap and so on. The second layer is the hidden layer comprising Neurons. The third layer is the output layer describing the output variables such as values of the strain predictions. The data are processed through the second and third layer with the activation function [29]. The relationship model between input set $\{X_m | m = 1, 2, \cdots, J\}$ and strain Y is established using an improved neural network algorithm. With samples $X_1, \cdots, X_m$ as the input values, and $Y_1, \cdots, Y_L$ as the output values, the strain prediction model is trained. The network architecture of BPNN predicting the wind turbine blade static test strain is illustrated by Figure 5.

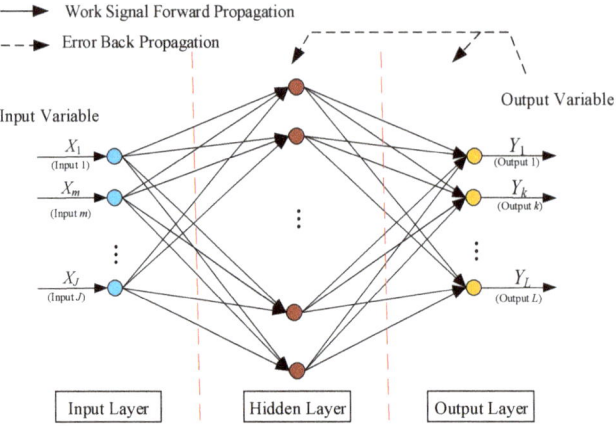

Figure 5. The topology of BPNN.

3.2. The BPNN Optimization Principles Based on PSO

The traditional BPNN utilizes the gradient descent method generally, so the PSO-BPNN in our research is improved by Levenberg-Marquart. The convergence of BPNN is mainly dependent on the initial weights and thresholds and it is likely to obtain the local minimum. According to these characteristics of the BPNN, it is a kind of popular and effective method to overcome the defects that the naturally inspired algorithms have [30]. In this research, the weight and threshold values of BPNN, which can effectively prevent the training from falling into the local optimal situation, are optimized by PSO, which means errors of this neural network training are minimized. In addition, PSO is easy to use and perform its function because it is not necessary to set parameters for evolving such as crossover operator and mutational operator [31]. The improved BPNN process based on PSO is shown in Figure 6. Many possible initial values can be set by PSO, which updates them in a range of continuous iterations. The different conditional combinations can be achieved in each iteration until the biggest fitness function value is obtained and the output values are computed. The optimization process of PSO-BPNN is shown in Figure 6.

PSO has good global search capability, accordingly updating the velocities and positions of particles, the global best particle is found. The weight and threshold of BPNN are optimized, then the BPNN is trained. After that, the optimization structure of BPNN is completed. The steps of BPNN improved by PSO are summarized as follows [32]:

First of all, the size of the population is set. The frontier $[-X_{\max}, X_{\max}]$, the biggest velocity $[-V_{\max}, V_{\max}]$, inertia weight coefficient w, the maximum iterative number, and the acceleration constants c_1, c_2 are set. The position X_i and velocity V_i are initialized.

Second, the fitness values of each particle are calculated. According the fitness function, all the fitness values are obtained. The higher the fitness values are, the better the performance in particles is. Meanwhile, the individual best positions P_{best} and the global optimum positions of particles G_{best} are updated and recorded.

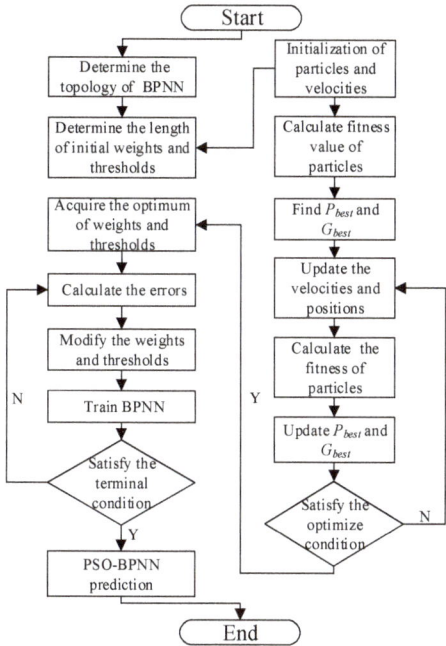

Figure 6. The optimization process of PSO-BPNN.

Third, the velocity and position of every particle updated can be obtained using Equations (1) and (2) [33]:

$$V_{ij}^{k+1} = wV_{ij}^k + c_1 r_1 \left(P_{ij}^k - X_{ij}^k\right) + c_2 r_2 \left(G_j^k - X_{ij}^k\right) \tag{1}$$

$$X_{ij}^{k+1} = X_{ij}^k + V_{ij}^{k+1}, 1 \le i \le n, 1 \le j \le d \tag{2}$$

where k represents discrete time index, w represents inertia weight factor, c_1 and c_2 represent acceleration constants, and r_1 and r_2 are uniformly distributed random numbers on the interval [0, 1]. P_{ij}^k is dimension j of the best point vector found by the i-th particle in the k-th iteration, and G_j^k is dimension j of the global best point in the k-th iteration. X_{ij}^k represents dimension j of the position vector found by the ith particle in the k-th iteration, and V_{ij}^{k+1} is dimension j of the velocity vector found by the i-th particle in the $(k + 1)$-th iteration.

Generally, if the velocities of the flying swarms are not restricted appropriately during the PSO calculation, the swarms may fly into the local optimization, which means the flying particles cannot reach the global optimum. For the purpose of the particles based on the restricted velocity to obtain the optimal solution, V_{max} as the velocity threshold is introduced, and the limited conditions are imposed as follows:

If $V_{ij}^{(k+1)} > V_{max}$, then $V_{ij}^{(k+1)} = V_{max}$;

If $V_{ij}^{(k+1)} < -V_{max}$, then $V_{ij}^{(k+1)} = -V_{max}$; else $V_{ij}^{(k+1)}$ unchanged.

If $X_{ij}^{(k+1)} > X_{max}$, then $X_{ij}^{(k+1)} = X_{max}$; If $X_{ij}^{(k+1)} < X_{min}$, then $X_{ij}^{(k+1)} = X_{min}$.

Fourth, if the algorithm reaches the maximum iteration or the precision of error is smaller than the setting value, the point of the current optimization swarm is outputted. Otherwise, this process should return to the third step and continue to train the PSO model.

Fifth, G_{best} outputted depending on the fourth step are as the initial inputted values of the weight and threshold of BPNN. In the PSO-BPNN algorithm, every dimension of the vector $x_i = (x_{i1}, x_{i2}, \cdots, x_{is})$ represents the weight or threshold values of BPNN; s is the amount of weights and thresholds in BPNN. The fitness functions of particle swarms are as Formulae (3) and (4) [32]:

$$L_i = \sum_{j=1}^{s} (y_{ij} - Y_{ij})^2 \tag{3}$$

$$L_{popIndex} = \frac{1}{m} \sum_{i=1}^{m} L_i \tag{4}$$

where m is the number of particles describing population size, Y_{ij} is the ideal outputted value j of the i-th particle, Y_{ij} is the actual outputted value j of the i-th particle; $popIndex = 1, 2, \cdots, m$. $L_{popIndex}$ is the fitness of particle $popIndex$.

Sixth, the established training set is applied to the optimized BPNN algorithm. The transfer function is the tangent S-type transfer function between the input layer and the hidden layer too, and the transfer function is also set as a linear transfer function between the hidden layer and the output layer.

4. Strain Prediction Model Based on PSO-BPNN for the Wind Turbine Blade Static Behaviour

4.1. Strain Prediction Modeling for the Central of Pressure Side on Edgewise+

When the wind turbine blade bears unilateral tension, the relationship among the strain and the applied loads, loading positions and displacements is nonlinear. The PSO-BPNN methods have outstanding advantages to characterize the nonlinear relationship, thus the strain values of the unmeasured points of the wind turbine blade can be predicted by the PSO-BPNN models. In the PSO-BPNN models for strain predicting, the values of the applied loads, the loading positions and the displacements are used as training inputs, and the strain values are outputted. The 51 sets of strain testing data of the center of the pressure side when loaded on edgewise+ were taken as training samples and five sets of strain testing data were used as test samples, thus a strain predictive PSO-BPNN model for the central of PS was established. The training samples and test samples of the PSO-BPNN model are shown in Tables 5 and 6, respectively.

Table 5. The training samples used by BPNN and PSO-BPNN model.

Items	Locations of Strain Gauges	Load Applications (kN)					
		F_1	F_2	F_3	F_4	F_5	F_6
1	2000	44.9	25.6	18.6	14.1	8.9	73.9
2	6000	66.8	37.7	27.9	20.9	13.4	71
3	12,000	89.4	50.7	36.1	27.7	17.7	68.2
...	...	...	...	...	...	...	...
50	45,000	112	62.9	45.4	34.5	22.1	65.4
51	33,000	44.9	25.6	18.6	14.1	8.9	73.9

Items	Displacements of the Loading Positions (mm)					Strain (με)	
	s_1	s_2	s_3	s_4	s_5	s_6	f
1	88	245	504	742	1093	1284	−55.2
2	135	379	736	1083	1594	1874	−85.4
3	188	507	973	1430	2104	2472	−140.1
...	...	...	...	...	...	...	...
50	240	635	1220	1794	2633	3091	−231.7
51	88	245	504	742	1093	1284	−140.2

The PSO-BPNN model is operated with the following settings: the maximum number of PSO iterations is 120; both the acceleration constants c_1 and c_2 are set to 2; the maximum particle velocity is

$0.8 \times X_{max}$ and the minimum particle velocity is $0.6 \times X_{min}$; the inertia weight is 0.2. The inputting dimensions are set to 13, and the outputting dimension is set to 1. There are 17 neuron nodes in the hidden layer of the network. The maximum number of network training epochs allowed is 5000; the speed of network learning is 0.1; the minimum convergence error of the training target is set to 0.001. Then, the learning procedure depending on the project samples was calculated. In order to verify the validity and superiority of the PSO-BPNN model, the BPNN algorithm is used solely for the strain predictions to compare with the PSO-BPNN model. The parameters settings of the BPNN algorithm running are shown as follows: The number of inputting dimensions is set to 13 in addition with the number of outputting dimension setting to 1; there are 17 neuron nodes set into the hidden layer, too; the maximum number of network training epochs allowed is set to 5000; the speed of network learning is set to 0.1; the minimum convergence error of the training target is set to 0.001; and the frequency of displaying results is set to every 50 steps. Then, the learning procedure of the project samples was operated. The results of comparisons calculated are illustrated in Figures 7 and 8.

Table 6. The test samples used by BPNN and PSO-BPNN model.

Items	Locations of Strain Gauges	Load Applications (kN)					
		F_1	F_2	F_3	F_4	F_5	F_6
1	36,000	89.4	50.7	36.1	27.7	17.7	68.2
2	9000	66.8	37.7	27.9	20.9	13.4	71
3	2100	89.4	50.7	36.1	27.7	17.7	68.2
4	12,000	66.8	37.7	27.9	20.9	13.4	71
5	15,000	44.9	25.6	18.6	14.1	8.9	73.9

Items	Displacements of Loading Positions (mm)						Strain (με)
	s_1	s_2	s_3	s_4	s_5	s_6	f
1	88	245	504	742	1093	1284	−71.1
2	135	379	736	1083	1594	1874	−91.5
3	188	507	973	1430	2104	2472	−271.2
4	135	379	736	1083	1594	1874	−106.9
5	88	245	504	742	1093	1284	−105.5

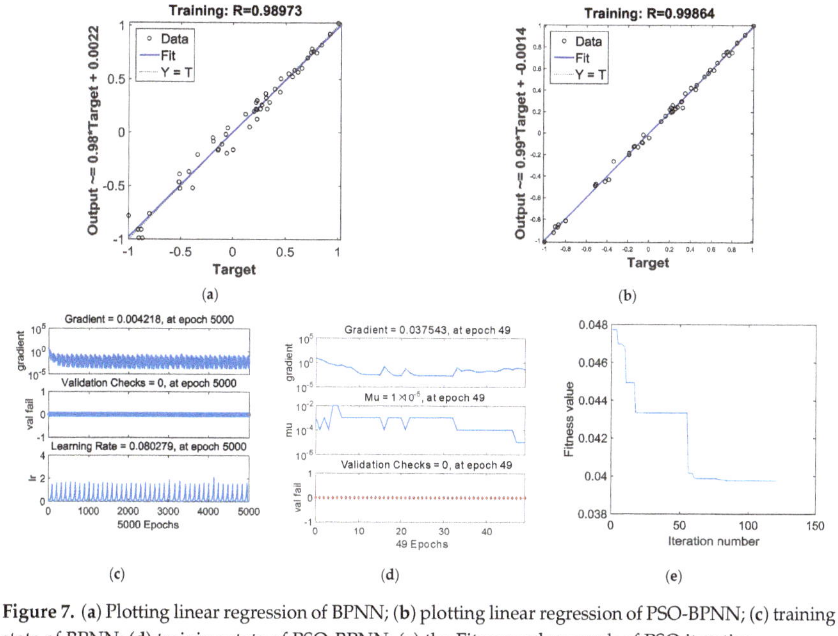

Figure 7. (**a**) Plotting linear regression of BPNN; (**b**) plotting linear regression of PSO-BPNN; (**c**) training state of BPNN; (**d**) training state of PSO-BPNN; (**e**) the Fitness value graph of PSO iteration.

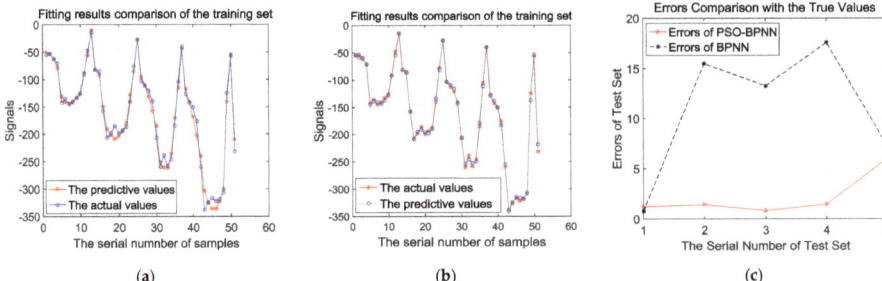

Figure 8. (a) Fitting results comparison of the training set by BPNN; (b) fitting results comparison of the training set by PSO-BPNN; (c) errors comparison of test set by BPNN and PSO-BPNN.

Figure 7a,b plots the linear regression of BPNN and PSO-BPNN respectively. The regression coefficient of PSO-BPNN, measuring the correlation between outputs and targets, is closer to 1 than that of BPNN, which means the PSO-BPNN training has better performance than the BPNN training. Figure 7c,d shows the training state of BPNN and PSO-BPNN, respectively; the PSO-BPNN ran 49 epochs whereas BPNN ran 5000 epochs to the ideal values, so the network trained by PSO-BPNN has higher efficiency than BPNN. In Figure 7e, the fitness values of PSO are shown. When the number of iteration times reaches the early 30s, the fitness values achieve the ideal results in the PSO process.

Figure 8a,b presents the comparison of fitting effects based on the training samples analyzed by BPNN and PSO-BPNN, respectively. They show that the fitting effects of PSO-BPNN are better compared to BPNN. Besides, in order to verify the fitting and predictive abilities of the PSO-BPNN and BPNN algorithms, the result of errors comparison with the true values is shown after the test samples are trained by the two different algorithms in Figure 8c. In Figure 8c, we can see that the PSO-BPNN model has much higher accuracy according to the test results, and its rates of relative errors outputted according to the test sample training are below 6%, while the relative error rates of BPNN models are below 18%, so PSO-BPNN shows a better performance of predictive ability than BPNN.

4.2. Strain Prediction Modeling for the Central of Pressure Side on Edgewise−

The methodology is the same as that used in Section 4.1. A strain-predictive PSO-BPNN model for the center of the pressure side loaded on the direction of edgewise− is established. The training samples and test samples of the strain-predictive PSO-BPNN model are shown in Tables 7 and 8, respectively.

Table 7. The training samples trained by BPNN and PSO-BPNN model.

Items	Locations of Strain Gauges	Load Applications (kN)				
		F_1	F_2	F_3	F_4	F_5
1	6000	19	32.7	12.3	9.6	9.6
2	12,000	27.6	48.9	18.7	14.2	14.5
3	21,000	36.8	65.2	24.3	19	19.2
...	...	...	...	...	...	...
...	...	...	...	...	...	...
51	51,000	46	81.7	30.5	23.6	24

Items	Displacements of the Loading Positions (mm)					Strain (με)
	s_1	s_2	s_3	s_4	s_5	s_6	f
1	66	198	419	616	912	1079	11.3
2	103	307	629	924	1368	1618	40.9
3	141	417	838	1230	1822	2152	140.8
...	...	...	...	...	...	...	...
...	...	...	...	...	...	...	...
51	181	528	1054	1544	2286	2697	76.9

Table 8. The test samples trained by BPNN and PSO-BPNN model.

Items	Locations of Strain Gauges	Load Applications (kN)				
		F_1	F_2	F_3	F_4	F_5
1	33,000	19	32.7	12.3	9.6	9.6
2	36,000	27.6	48.9	18.7	14.2	14.5
3	27,000	27.6	48.9	18.7	14.2	14.5
4	39,000	36.8	65.2	24.3	19	19.2
5	45,000	36.8	65.2	24.3	19	19.2

Items	Displacements of the Loading Positions (mm)						Strain (με)
	s_1	s_2	s_3	s_4	s_5	s_6	f
1	66	198	419	616	912	1079	34.6
2	103	307	629	924	1368	1618	52.9
3	103	307	629	924	1368	1618	70.5
4	141	417	838	1230	1822	2152	70.2
5	141	417	838	1230	1822	2152	54.9

Figure 9a,b shows the linear regressions of BPNN and PSO-BPNN respectively; the regression R-value of the PSO-BPNN model training is 0.99971, which is also closer to 1. It means that the training results of PSO-BPNN show closer correlation with targets than that of BPNN, so the performance of PSO-BPNN training is better than that of BPNN algorithm. Figure 9c,d shows the training state of BPNN and PSO-BPNN, respectively. Figure 9e presents the fitness value graph of PSO. The fitness value levels off between the 165th and the 175th iteration time.

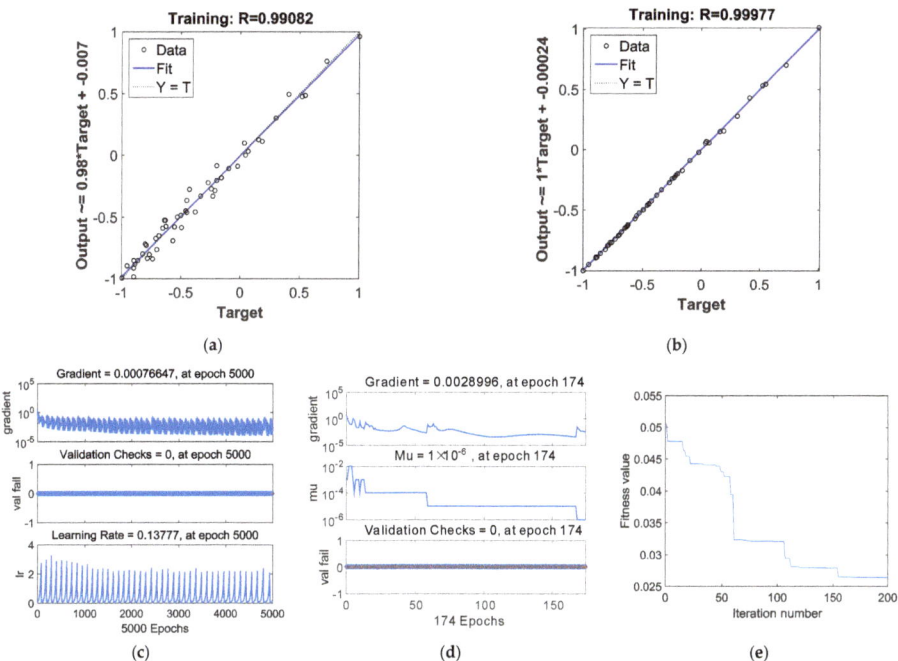

Figure 9. (a) Plotting linear regression of BPNN; (b) plotting linear regression of PSO-BPNN; (c) training state of BPNN; (d) training state of PSO-BPNN; (e) the Fitness value graph of PSO iteration.

According to Figure 10a,b, the fitting figure of the set trained by PSO-BPNN is more precise than that set trained by BPNN. In addition, the test samples are utilized to prove the recognition ability of the trained PSO-BPNN, and the comparison results are shown in Figure 10c. In Figure 10c, we can find that test results of the PSO-BPNN model are much more accurate, the average error rate of PSO-BPNN is less than that of BPNN, and all of the error rates regarding test samples output are below 5.8%, while the relative error rates analyzed by BPNN are all within 6%.

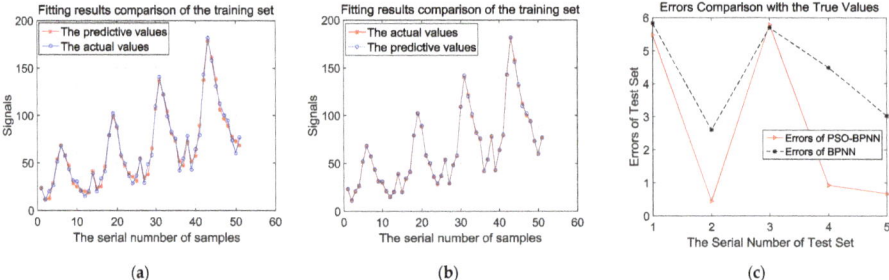

Figure 10. (a) Fitting results comparison of the training set by BPNN; (b) fitting results comparison of the training set by PSO-BPNN; (c) test set errors comparison of BPNN and PSO-BPNN.

4.3. Predicted Results and Verification

In order to demonstrate the effectiveness and feasibility of the proposed PSO-BPNN strain-predictive method, the prediction results based on BPNN methods are used to compare with the FE simulation results. For unmeasured points on the center of the pressure side, 17 points located at 11.00 m, 14.00 m, 16.00 m, 19.00 m, 20.00 m, 20.64 m, 23.00 m, 25.00 m, 33.00 m, 35.00 m, 38.00 m, 41.00 m, 42.00 m, 43.00 m, 46.00 m, 49.00 m, and 52.00 m are chosen to predict their strain value by PSO-BPNN and BPNN methods.

When loaded on edgewise+, the prediction strain values of the 17 unmeasured points are shown in Figure 11. From Figure 11, we can see that the strain value decreases a lot at the early stage, stays flat for a period of time, and then increases. This result is in accordance with the structural characteristics of blades, owning to a reinforced structure set near the maximum chord length. Besides, the predicted results of all BPNN models are very close to the simulation results, and that means BPNN strain-predictive methods have a high accuracy to predict the strain. Compared with the traditional BPNN method, the PSO-BPNN method has the smallest error, thus the strain-predictive model based on PSO-BPNN is scientific and reasonable.

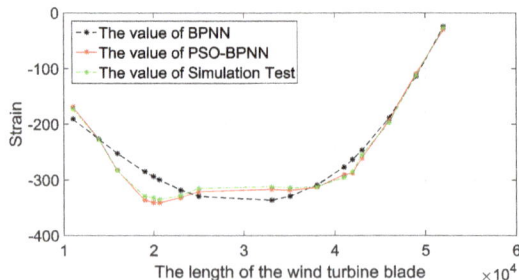

Figure 11. Comparisons of BPNN methods with the FEA.

When loaded on edgewise−, the strain values of the above 17 unpredicted points are shown in Figure 12. The conclusion is the same as the analysis results: When loaded on the direction of edgewise+, all strain-predictive BPNN models have high accuracies, and the PSO-BPNN has the smallest error.

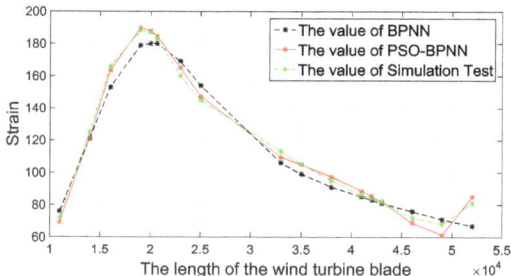

Figure 12. Comparisons of BPNN methods with the FEA.

According to the prediction result of comparisons among BPNN models and the FE simulation results, all BPNN models can predict the strain effectively, while the PSO-BPNN method has higher accuracies compared with traditional BPNN method, and it is more suitable to predict the strain of unmeasured points in the full-scale static testing of wind turbine blades.

5. Conclusions

In the full-scale static testing of wind turbine blades, the correlation among the strain and applied loads, loading positions, displacements, etc., is nonlinear, and the number of design variables is numerous, thus the calculation and prediction of the blade strain are very complicated and difficult by traditional numerical methods. Considering these reasons, a strain-predictive PSO-BPNN method is proposed:

(1) Taking the advantages of the BPNN methods in dealing with the nonlinear relationship, a strain-predictive PSO-BPNN model for the full-scale static testing of wind turbine blades was established;

(2) The accuracy of the strain-predictive PSO-BPNN model was verified by comparisons with the traditional BPNN models as well as the ANSYS simulation test. When loaded on the direction of edgewise+, the relative error rate of the strain-predictive PSO-BPNN model is within 6%. Similarly, when loaded on the direction of edgewise−, the relative error rate is also within 6%, which satisfies the blade certification requirements.

(3) The applicability and usability of the strain-predictive BPNN models were verified by comparing with the AYSYS simulation test for 17 unmeasured points. From the comparison results, we can see that all BPNN models have high accuracies to predict strains, and the PSO-BPNN method has the smallest error. Thus, the PSO-BPNN method is much more suitable to predict the strain of unmeasured points in the full-scale static testing of wind turbine blades.

A strain-predictive PSO-BPNN model for full-scale static testing of the wind turbine blade was established in this paper and more strain values can be predicted for unmeasured points in the full-scale static testing of wind turbine blades. This study can provide more data to verify the rationality of blade design and correct the blade defects; the outputs can also be used for life prediction for wind blades, which will be considered as the next work. Moreover, the number of test samples is chosen on the basis of the static test, while the relation between sample number and accuracy will also be considered and studied in the future.

Author Contributions: Conceptualization, X.L., Z.L.(Zheng Liu), J.A.F.O.C. and A.M.P.D.J.; methodology, Z.L. (Zheng Liu), X.L., J.A.F.O.C. and A.M.P.D.J.; validation, Z.L.(Zheng Liu) and Z.L. (Zhongwei Liang); formal analysis, X.L., Z.L.(Zheng Liu), S.-P.Z.; investigation, Z.L. (Zheng Liu), S.-P.Z.; data curation, X.L., Z.L. (Zheng Liu); Writing-original draft preparation, X.L., Z.L. (Zheng Liu), Z.L. (Zhongwei Liang), S.-P.Z.; writing—review and editing, J.A.F.O.C. and A.M.P.D.J.; supervision, Z.L. (Zheng Liu)and J.A.F.O.C.; project administration, J.A.F.O.C.; funding acquisition, Z.L. (Zheng Liu), X.L., Z.L. (Zhongwei Liang) and J.A.F.O.C.

Funding: This research was funded by Science and Technology Program of Guangzhou, China (201904010463), Guangzhou University Teaching Reform Project (09-18ZX0309), the Guangzhou University Teaching Reform Project (09-18ZX0304), the Innovative Team Project of Guangdong Universities (2017KCXTD025), and the Innovative Academic Team Project of Guangzhou Education System (1201610013).

Acknowledgments: The paper was partially supported by Science and Technology Program of Guangzhou, China (201904010463), Teaching Reform Project from Guangzhou University (09-18ZX0309), Teaching Reform Project from Guangzhou University (09-18ZX0304), Science and Technology Fund of Guangzhou City (201504291326362), the Innovative Team Project of Guangdong Universities (2017KCXTD025), and the Innovative Academic Team Project of Guangzhou Education System (1201610013). This work was also supported by national funds through FCT—Fundação para a Ciência e a Tecnologia; UID/ECI/04708/2019—CONSTRUCT—Instituto de I&D em Estruturas e Construções funded by national funds through the FCT/MCTES (PIDDAC).

Conflicts of Interest: The authors declare no conflict of interest.

Abbreviations

BPNN	Back Propagation Neural Network
PSO	Particle Swarm Optimization
PSO-BPNN	Back Propagation Neural Network Improved by Particle Swarm Optimization
PS	Pressure Side
SS	Suction Side

References

1. Veritas, D.N. *Design and Manufacture of Wind Turbine Blades, Offshore and Onshore Wind Turbines*; Det Norske Veritas: Oslo, Norway, 2010.
2. Samborsky, D.; Mandell, J.; Sears, A.; Kils, O. Static and Fatigue Testing of Thick Adhesive Joints for Wind Turbine Blades. *ASME Wind Energy Symp.* **2009**.
3. Jensen, F.M.; Falzon, B.G.; Ankersen, J.; Stang, H. Structural testing and numerical simulation of a 34m composite wind turbine blade. *Compos. Struct.* **2006**, *76*, 52–61. [CrossRef]
4. Lee, H.G.; Kang, M.G.; Park, J. Fatigue failure of a composite wind turbine blade at its root end. *Compos. Struct.* **2015**, *133*, 878–885. [CrossRef]
5. Pan, Z.J.; Wu, J.Z.; Liu, J.; Zhao, X.H. Fatigue failure of a composite wind turbine blade at the trailing edge. *Defect Diffus. Forum* **2018**, *382*, 191–195. [CrossRef]
6. Lee, H.G.; Park, J. Static test until structural collapse after fatigue testing of a full-scale wind turbine blade. *Compos. Struct.* **2016**, *136*, 251–257. [CrossRef]
7. Zhu, S.P.; Yu, Z.Y.; Liu, Q.; Ince, A. Strain energy-based multiaxial fatigue life prediction under normal-shear stress interaction. *Int. J. Damage Mech.* **2019**, *28*, 708–739. [CrossRef]
8. Ai, Y.; Zhu, S.P.; Liao, D.; Correia, J.A.F.O.; Souto, C.; De Jesus, A.M.P.; Keshtegar, B. Probabilistic modeling of fatigue life distribution and size effect of components with random defects. *Int. J. Fatigue* **2019**, *126*, 165–173. [CrossRef]
9. Ai, Y.; Zhu, S.P.; Liao, D.; Correia, J.A.; De Jesus, A.M.; Keshtegar, B. Probabilistic modelling of notch fatigue and size effect of components using highly stressed volume approach. *Int. J. Fatigue* **2019**, *127*, 110–119. [CrossRef]
10. Choi, K.S.; Huh, Y.H.; Kwon, I.B.; Yoon, D.J. A tip deflection calculation method for a wind turbine blade using temperature compensated FBG sensors. *Smart Mater. Struct.* **2012**, *21*, 025008. [CrossRef]
11. Kim, S.W.; Kang, W.R.; Jeong, M.S.; Lee, I.; Kwon, I.B. Deflection estimation of a wind turbine blade using FBG sensors embedded in the blade bonding line. *Smart Mater. Struct.* **2013**, *22*, 125004. [CrossRef]
12. Roczek-Sieradzan, A.; Nielsen, M.; Branner, K.; Jensen, F.M.; Bitsche, R. Wind turbine blade testing under combined loading. *Proceed. Risø Int. Symp. Mater. Sci.* **2011**, *32*, 449–456.
13. Dou, H.Y.; Zhou, H.F.; Qin, Z.L.; Xie, Z.L.; Hao, S.W. Test and analysis of full-field 3D deformation for a wind turbine blade. *Acta Energ. Sol. Sin.* **2015**, *36*, 2257–2262.
14. Shi, K.Z.; Zhao, X.L.; Xu, J.Z. Research on fatigue test of large horizontal axis wind turbine blade. *Acta Energ. Sol. Sin.* **2011**, *32*, 1264–1268.
15. Pan, Z.J.; Wu, J.Z. Effects of structure nonlinear on full-scale wind turbine blade static test. *J. Tongji Univ. (Nat. Sci.)* **2017**, *45*, 1491–1497.

16. Yan, W.J.; Han, X.Y.; Cheng, L. Structure analysis and test of large-scale wind turbine blade. *Renew. Energy Resour.* **2014**, *32*, 1140.
17. Liao, D.; Zhu, S.P.; Correia, J.A.F.O.; de Jesus, A.M.P.; Calçada, R. Computational framework for multiaxial fatigue life prediction of compressor discs considering notch effects. *Eng. Fract. Mech.* **2018**, *202*, 423–435. [CrossRef]
18. Meng, D.; Yang, S.; Zhang, Y.; Zhu, S.P. Structural reliability analysis and uncertainties-based collaborative design and optimization of turbine blades using surrogate model. *Fatigue Fract. Eng. Mater. Struct.* **2019**, *42*, 1219–1227. [CrossRef]
19. Zhu, S.P.; Yue, P.; Yu, Z.Y.; Wang, Q. A combined high and low cycle fatigue model for life prediction of turbine blades. *Materials* **2017**, *10*, 698. [CrossRef]
20. Zhu, S.P.; Liu, Q.; Peng, W.; Zhang, X.C. Computational-experimental approaches for fatigue reliability assessment of turbine bladed disks. *Int. J. Mech. Sci.* **2018**, *142–143*, 502–517. [CrossRef]
21. Meng, D.; Liu, M.; Yang, S.; Zhang, H.; Ding, R. A fluid-structure analysis approach and its application in the uncertainty-based multidisciplinary design and optimization for blades. *Adv. Mech. Eng.* **2018**, *10*, 1–7. [CrossRef]
22. Zhu, S.P.; Liu, Y.; Liu, Q.; Yu, Z.Y. Strain energy gradient-based LCF life prediction of turbine discs using critical distance concept. *Int. J. Fatigue* **2018**, *113*, 33–42. [CrossRef]
23. Lin, W.; Liu, X.; Renevier, N.; Stables, M.; Hall, G.M. Nonlinear aeroelastic modelling for wind turbine blades based on blade element momentum theory and geometrically exact beam theory. *Energy* **2014**, *76*, 487–501.
24. Kusiak, A.; Zhang, Z. Adaptive Control of a Wind Turbine with Data Mining and Swarm Intelligence. *IEEE Trans. Sustain. Energy* **2011**, *2*, 28–36. [CrossRef]
25. Cynthia, J.E.J.; Darwin, J.D.; Jeyanthy, P.A.; Darwin, J.D.; Devika, S. Power Coefficient in Wind Power Using Particle Swarm Optimization. In Proceedings of the 2014 International Conference on Control, Instrumentation, Communication and Computational Technologies (ICCICCT), Kanyakumari, India, 10–11 July 2014.
26. Fooladi, M.; Akbari Foroud, A. Recognition and assessment of different factors which affect flicker in wind turbines. *IET Renew. Power Genera.* **2016**, *10*, 250–259. [CrossRef]
27. Wang, L.; Hu, P.; Lu, J.; Chen, F.; Hua, Q. Neural network and PSO-based structural approximation analysis for blade of wind turbine. *Int. J. Modell. Identif. Control* **2013**, *18*, 69–75. [CrossRef]
28. Lin, S.; Chen, S.; Wu, W.; Chen, C.H. Parameter determination and feature selection for back-propagation network by particle swarm optimization. *Knowl. Inf. Syst.* **2009**, *21*, 249–266. [CrossRef]
29. Hossain Lipu, M.S.; Hannan, M.A.; Hussain, A.; Saad, M.H. Optimal BP neural network algorithm for state of charge estimation of lithium-ion battery using PSO with PCA feature selection. *J. Renew. Sustain. Energy* **2017**, *9*, 64102. [CrossRef]
30. Nait Amar, M.; Zeraibi, N.; Redouane, K. Bottom hole pressure estimation using hybridization neural networks and grey wolves optimization. *Petroleum* **2018**, *4*, 419–429. [CrossRef]
31. Nayak, D.R.; Dash, R.; Majhi, B. Discrete ripplet-II transform and modifie d PSO based improved evolutionary extreme learning machine for pathological brain detection. *Neurocomputing* **2018**, *282*, 232–247. [CrossRef]
32. Jin, C.; Jin, S.; Qin, L. Attribute selection method based on a hybrid BPNN and PSO algorithms. *Appl. Soft Comput. J.* **2012**, *12*, 2147–2155. [CrossRef]
33. Kshirsagar, P.; Akojwar, S. Optimization of BPNN parameters using PSO for EEG signals. *Adv. Intell. Syst. Res.* **2017**, *137*, 385–394.

© 2019 by the authors. Licensee MDPI, Basel, Switzerland. This article is an open access article distributed under the terms and conditions of the Creative Commons Attribution (CC BY) license (http://creativecommons.org/licenses/by/4.0/).

Article

Reliability-Based Low Fatigue Life Analysis of Turbine Blisk with Generalized Regression Extreme Neural Network Method

Chunyi Zhang [1], Jingshan Wei [1], Huizhe Jing [1], Chengwei Fei [2,*] and Wenzhong Tang [3]

1. School of Mechanical and Power Engineering, Harbin University of Science and Technology, Harbin 150080, China; zhangchunyi@hrbust.edu.cn (C.Z.); wjs19931208@163.com (J.W.); jinghuizhe@163.com (H.J.)
2. Department of Aeronautics and Astronautics, Fudan University, Shanghai 200433, China
3. School of Computer Science and Technology, Beihang University, Beijing 10191, China; tangwenzhong@buaa.edu.cn
* Correspondence: cwfei@fudan.edu.cn

Received: 23 March 2019; Accepted: 8 May 2019; Published: 10 May 2019

Abstract: Turbine blisk low cycle fatigue (LCF) is affected by various factors such as heat load, structural load, operation parameters and material parameters; it seriously influences the reliability and performance of the blisk and aeroengine. To study the influence of thermal-structural coupling on the reliability of blisk LCF life, the generalized regression extreme neural network (GRENN) method was proposed by integrating the basic thoughts of generalized regression neural network (GRNN) and the extreme response surface method (ERSM). The mathematical model of the developed GRENN method was first established in respect of the LCF life model and the ERSM model. The method and procedure for reliability and sensitivity analysis based on the GRENN model were discussed. Next, the reliability and sensitivity analyses of blisk LCF life were performed utilizing the GRENN method under a thermal-structural interaction by regarding the randomness of gas temperature, rotation speed, material parameters, LCF performance parameters and the minimum fatigue life point of the objective of study. The analytical results reveal that the reliability degree was 0.99848 and the fatigue life is 9419 cycles for blisk LCF life when the allowable value is 6000 cycles so that the blisk has some life margin relative to 4500 cycles in the deterministic analysis. In comparison with ERSM, the computing time and precision of the proposed GRENN under 10,000 simulations is 1.311 s and 99.95%. This is improved by 15.18% in computational efficiency and 1.39% in accuracy, respectively. Moreover, high efficiency and high precision of the developed GRENN become more obvious with the increasing number of simulations. In light of the sensitivity analysis, the fatigue ductility index and temperature are the key factors of determining blisk LCF life because their effect probabilities reach 41% and 26%, respectively. Material density, rotor speed, the fatigue ductility coefficient, the fatigue strength coefficient and the fatigue ductility index are also significant parameters for LCF life. Poisson's ratio and elastic modulus of materials have little effect. The efforts of this paper validate the feasibility and validity of GRENN in the reliability analysis of blisk LCF life and give the influence degrees of various random parameters on blisk LCF life, which are promising to provide useful insights for the probabilistic optimization of turbine blisk LCF life.

Keywords: turbine blisk; low cycle fatigue life; reliability analysis; generalized regression neural network; extremum response surface method

1. Introduction

As a heat-end core component of an aeroengine, a turbine blisk endures complex alternating loads due to operation in a severe environment with high temperatures and high rotation speeds.

In this case, it is easy to produce large plastic deformation for blisk and to induce the low cycle fatigue (LCF) failure of blisk [1,2]. Most of the parameters that significantly effect blisk LCF failure have some randomness [3]. To improve the safety and reliability of a turbine blisk to ensure the high performance of an aeroengine, it is important to study blisk LCF life reliability from a probabilistic perspective [4–8].

The LCF life of structures has been widely investigated. Sun et al. established a nonlinear model for LCF life of a steam turbine rotor under a temperature-stress coupling field by considering the relationship between cyclic stress and strain and validated the model to be accurate and reasonable in describing damage accumulation [9]. Letcher et al. proposed an energy-based critical fatigue life prediction approach, which derived the approximate failure cycle index from the ratio of the total accumulation of energy in the fracture process to the one-cycle strain energy [10]. Bargmann et al. discussed the full-probability quick integral algorithm based on the Coffin-Manson-Neuber local strain-fatigue theory [11]. Zhu et al. discussed the probabilistic LCF life prediction of a turbine disk under uncertainties [12,13]. Viadro et al. studied the reliability of stiffened bending plates [14]. Repetto et al. discussed the role of parameter uncertainty in the damage prediction of the alongwind-induced fatigue and long term simulation of wind-induced fatigue loadings [15,16]. Most of the above work was conducted based on numerical simulation methods (or-called direct simulation methods) with Monte Carlo (MC) simulation [15–20]. Generally, the direct simulation methods are powerful for the deterministic analyses of component LCF life. However, for the probabilistic analyses of component LCF life with thousands of iterations and MC simulations, it is unbelievable to efficiently perform blisk fatigue life analysis owing to excess computational burden (loads) and unacceptable computational efficiency; although this method has satisfactory computing precision against engineering practice. Therefore, it is urgent to seek an alternative effective method for direct methods to address this issue.

In respect of the in-depth investigation of structural fatigue probabilistic analyses, the response surface method (RSM, also called surrogate model method) is indeed an alternative method to direct simulation methods [21–26]. With the development of structural reliability theory and methods, various surrogate methods have emerged [27–29]. To improve the computational efficiency and accuracy of RSM for complex structural reliability analysis, Bai et al. proposed a distributed collaborative response surface method for the mechanical dynamic assemble reliability analysis of aeroengine high pressure turbine blade-tip clearance [30]. Hurtado et al. proposed a highly efficient surrogate method, a support vector machine, for structural reliability analysis with small samples [31]. Zhang et al. developed an extremum response surface method (ERSM) in respect of the extreme thought, to address the transient problem in the dynamic reliability analysis of a flexible mechanism and validated the ERSM to be precise for the reliability analysis of a flexible manipulator [32]. Lu et al. developed an improved Kriging method by integrating the Kriging algorithm and ERSM for the reliability and sensitivity analyses of a compressor blisk regarding multiple failure (deformation failure, stress failure and strain failure) modes [33]. From the efforts in References [32,33], it can be seen that ERSM has the potential to handle the transient problem in structural dynamic probabilistic analyses with a high simulation accuracy and efficiency, which provides useful insight into the process of the reliability analysis of blisk LCF life with the consideration of aeroengine operating conditions. For another, the developed ERSM does not satisfy the requirement of engineering in computing precision, derived from the weakness in processing the involved nonlinear probabilistic analyses.

With the development of neural network technology recently, the nonlinear problem was skillfully addressed by developing a generalized regression neural network (GRNN) due to the strong nonlinear mapping capability and robustness [34]. Zhao et al. established an all-purpose regression neural network model based on a freight volume condition and validated the effectiveness of this model in freight volume prediction by modeling adaptive training and extrapolation evaluation in terms of historical statistical data of freight volume and related samples and economic indicators [35]. Li et al. fused the drosophila optimization algorithm and GRNN to build the prediction model of power loads for power load prediction and this model had a strong nonlinear fitting ability [36]. Sun et al. compared the GRNN model with the back propagation neural network (BPNN) model based on air quality prediction and the GRNN method needed less training time and had better stability, a higher fitting precision as well, compared to the BPNN model [37]. Wang et al. validated the strengths of the GRNN method again by waveguide orientation [38]. Therefore, the GRNN method has been comprehensively verified to be highly computationally precise and efficient.

To effectively perform the reliability and sensitivity analyses of a turbine blsik LCF life, the generalized regression extremum neural network (GRENN) method is proposed in this paper; by integrating the transient procession ability of ERSM and the nonlinear mapping and small samples of GRNN, to collectively ensure and improve computing precision and efficiency. The reliability analysis of a turbine blisk LCF life was implemented based on the the developed GRENN, by considering random input variables of temperature, rotation speed, material parameters (density, Poisson's ratio and elastic modulus) and fatigue performance parameters (fatigue ductility coefficient, fatigue strength coefficient and fatigue ductility index and fatigue strength index) as well as the output response of the minimum fatigue life. The developed GRENN method was validated by comparison with the MC method and ERSM.

2. Basic Theory

2.1. Mathematical Model of Low Cycle Fatigue Life

The Mason-Coffin equation indicates the strain-fatigue life equation, which expresses the relationship between strain and the fatigue life of materials [39], i.e.,

$$\frac{\Delta \varepsilon}{2} = \frac{\sigma_f'}{E}(2N_f)^b + \varepsilon_f'(2N_f)^c \tag{1}$$

where $\Delta \varepsilon$ is the total strain of specific structure; E indicates the elasticity modulus; σ_f' denotes the fatigue strength coefficient; ε_f' is the fatigue ductility coefficient; b indicates the fatigue strength exponent; c stands for the fatigue ductility exponent; N_f expresses the LCF life.

Considering the mean stress, σ_m, inducted by complex loads during aeroengine operation, Equation (1) can be rewritten by the Morrow correction equation [40], i.e.,

$$\frac{\Delta \varepsilon}{2} = \frac{\sigma_f' - \sigma_m}{E}(2N_f)^b + \varepsilon_f'(2N_f)^c. \tag{2}$$

Considering many cyclic loads, the LCF life can be classically computed by the line damage accumulation (Miner) law, i.e.,

$$D = \sum_{i=1}^{r} \frac{n_i}{N_i} \tag{3}$$

in which D indicates the fatigue damage; r is the number of loading levels; n_i denotes the cyclic number under the ith loading level; N_i is the fatigue life corresponding to the ith loading level.

2.2. Mathematical Model of Extremum Response Surface Method

To effectively process the transient problem in the dynamic reliability analysis of blisk LCF life involving nonlinear and transient features of numerous parameters, i.e., gas temperature, rotation speed, material parameters (density, Posion's ratio, elasticity modulus) and fatigue performance parameters (fatigue strength coefficient, fatigue ductility coefficient, fatigue strength exponent and fatigue ductility exponent), the ERSM proposed in Reference [32] was adapted by simplifying the response process of the LCF life as an extreme value (maximum value or minimum value) in an analytical time domain.

When X and y_e were used to indicate the input parameters set and the output extremum response, the ERSM model $y_e(X)$ of the dynamic system [32] can be written as

$$y_e(X) = f(X) = \left\{ y_e^{(j)}\left(X^{(j)}\right) \right\} \quad (4)$$

where $X^{(j)}$ is the jth group of the input samples; $y_e^{(j)}\left(X^{(j)}\right)$ indicates the output extremum response during a time domain.

In previous studies, most of the ERSM models were built based on polynomials [33] and these models are usually inefficient in model fitting because the polynomials are unworkable for highly nonlinear problems and the large computing burden (requiring a large number of samples for modeling) required for blisk fatigue life probabilistic analysis. Thus, the GRNN method, with a strong nonlinear mapping ability and robustness, was applied by combining ERSM in this paper to address the issues of modeling precision and efficiency, which result from nonlinearity and large samples.

2.3. Mathematical Model of Generated Regression Extremum Neural Network Method

GRNN is a feedforward neural network model based on nonlinear regression theory, including input layer (first layer), hidden layer (middle layer) and output layer (last layer), as shown in Figure 1.

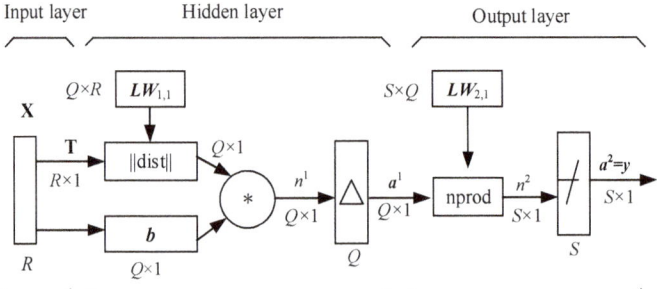

Note: X—matrix of input samples; T—matrix of output samples; Q—number of training samples; R—dimension number of input parameters; S—dimension number of output parameters; $LW_{1,1}$—weighted matrix in hide layer; $Q \times R$—dimensions of matrix $LW_{1,1}$; $||\text{dist}||$—weight (Euclidean distance) function in hide layer; n^1—network vector in hide layer; $\triangle$—transfer function in hide layer (Gauss function used in this paper); a^1—output of neuro cell in hide layer; $LW_{2,1}$—connection threshold value between hide layer and output layer; $S \times Q$—dimensions of matrix $LW_{2,1}$; nprod—weight function of output layer; n^2—network vector of output layer; $ブ$—linear transfer function (purelin) of output layer; a^2—output of neuro cell in output layer; y—output of neuro network

Figure 1. Schematic diagram of generalized regression extreme neural network (GRNN) method.

By inputting train samples into the input layer, the input matrix X and output matrix (denoted by T) are expressed by

$$X = \begin{bmatrix} x_{11} & x_{12} & \cdots & x_{1Q} \\ x_{21} & x_{22} & \cdots & x_{2Q} \\ \cdots & \cdots & \cdots & \cdots \\ x_{R1} & x_{R2} & \cdots & x_{RQ} \end{bmatrix}, T = \begin{bmatrix} t_{11} & t_{12} & \cdots & t_{1Q} \\ t_{21} & t_{22} & \cdots & t_{2Q} \\ \cdots & \cdots & \cdots & \cdots \\ t_{S1} & t_{S2} & \cdots & t_{SQ} \end{bmatrix} \quad (5)$$

where x_{ji} ($i = 1,2, \ldots, Q; j = 1,2, \ldots, R$) is the jth input sample in the ith group of training samples; t_{ji} indicates the jth output sample in the ith group of training samples; R is the number of input variables; S is the number of output variables; Q is the number of samples in the training set.

The number of neurons in the hidden layer was equal to the number of samples in the training set, the layer weight function was the Euclidean distance function (expressed in ||dist||), and the implicit layer weight matrix is calculated as follows:

$$LW_{1,1} = X^T \tag{6}$$

The threshold for Q hidden layer neural units is b:

$$b = [b_1, b_2, \cdots b_Q]^T \tag{7}$$

in which $b_1 = b_1 = \cdots = b_Q = \frac{0.8326}{\sigma}$, σ is the smooth factor of the Gauss function.

The transfer function of the hidden layer is usually based on the Gaussian radial basis function. The number of neurons in the hidden layer Q is equal to the number of training samples and each neuron corresponds to one training sample. When the weight matrix and threshold value of the hidden layer neural unit were determined, the output a_i^j of the ith hidden layer neuron is

$$a_i^j = exp(-\frac{0.8326}{\sigma}||LW_{1,i} - x_j||^2), \; j = 1, 2, \cdots, Q; i = 1, 2, \cdots, Q \tag{8}$$

in which $LW_{1,i} = [x_{h1}, x_{h2}, \cdots, x_{hR}]^T$ (h=1,2, $\cdots$, Q) is the vector of the ith implicit layer weight matrix $LW_{1,1}$; $x_j = [x_{j1}, x_{j2}, \cdots, x_{jR}]^T$ is the vector of the jth training samples. let $a^j = [a_1^j, a_2^j, \cdots, a_i^j, \cdots a_Q^j]$, which is the output vector of Q nerve cells corresponding to the jth group of input samples.

Regarding the connection weight $LW_{2,1}$ between the hidden layer and output layer as the output matrix of the training set of samples, which is denoted by T, i.e.,

$$LW_{2,1} = T. \tag{9}$$

The output layer is the third layer of GRNN. Based on GRNN Equations (8) and (9), vector n^j can be computed by

$$n^j = \frac{LW_{2,1}[a^j]^T}{\sum_{i=1}^{Q} a_i^j}. \tag{10}$$

With regard to the line transfer function $y^j = purelin(n^j)$ of n^j, the mathematical model of GRNN for the response of the jth group of training samples is expressed by

$$y^j = purelin(n^j) = \frac{LW_{2,1}[a^j]^T}{\sum_{i=1}^{Q} a_i^j} \tag{11}$$

where exp is a natural exponential function.

With respect to the format in Equation (4), the mathematical model of GRENN is

$$y_{min}^j = Min\left\{\frac{LW_{2,1}[a^j]^T}{\sum_{i=1}^{Q} a_i^j}\right\}. \tag{12}$$

2.4. Reliability Sensitivity Analyses Approaches with GRENN Model

Assuming that y^* is the allowable LCF life and y_{min}^j is the performance function of the structural fatigue life, the limit state function of LCF life is derived as [24]

$$Z = y_{min}^j - y^*. \tag{13}$$

In Equation (13), Z > 0 indicates that the blisk structure is secure, while Z < 0 reveals a failure. When random input variables are independently mutual, their means and variance are denoted by $\mu = [\mu_1, \mu_2 \cdots \mu_n]$ and $D = [D_1, D_2 \cdots D_n]$, respectively, we can gain [18]

$$\begin{cases} E(Z) = \mu_Z(\mu_1, \mu_2, \cdots, \mu_n; D_1, D_2, \cdots, D_n) \\ D(Z) = D_Z(\mu_1, \mu_2, \cdots, \mu_n; D_1, D_2, \cdots, D_n) \end{cases} \quad (14)$$

in which $E(Z)$ is mean function and $D(Z)$ is variance function.

When the limit state function of the structural LCF life (Equation (13)) obeys a normal distribution, the reliability degree P_r is expressed as [25]

$$P_r = \Phi\left(\frac{\mu_Z}{\sqrt{D_Z}}\right) \quad (15)$$

where μ_z is the mean matrix of a limit state function Z; D_z is the variance matrix of a limit state function.

The sensitivity reflects the level of sensitivity of the input random variables on the failure probability of a structural system response, which is promising to determine the extent to which these parameters effect the response and then provide a useful guide for structural design and optimization [41].

With the proposed GRENN method, the sensitivity degree can be determined by the mean matrix μ and variance D of input random variables [42], i.e.,

$$\frac{\partial P_r}{\partial \mu^T} = \frac{\partial P_r}{\partial \left(\frac{\mu_Z}{\sqrt{D_Z}}\right)} \left(\frac{\partial \left(\frac{\mu_Z}{\sqrt{D_Z}}\right)}{\partial \mu_Z} \frac{\partial \mu_Z}{\partial \mu^T} + \frac{\partial \left(\frac{\mu_Z}{\sqrt{D_Z}}\right)}{\partial D_Z} \frac{\partial \mu_Z}{\partial \mu^T}\right); \frac{\partial P_r}{\partial D^T} = \frac{\partial P_r}{\partial \left(\frac{\mu_Z}{\sqrt{D_Z}}\right)} \left(\frac{\partial \left(\frac{\mu_Z}{\sqrt{D_Z}}\right)}{\partial \mu_Z} \frac{\partial \mu_Z}{\partial D^T} + \frac{\partial \left(\frac{\mu_Z}{\sqrt{D_Z}}\right)}{\partial D_Z} \frac{\partial \mu_Z}{\partial D^T}\right) \quad (16)$$

in which

$$\begin{cases} \frac{\partial P_r}{\partial (\mu_Z/\sqrt{D_Z})} = P_r, \frac{\partial (\mu_Z/\sqrt{D_Z})}{\partial \mu_Z} = \frac{1}{\sqrt{D_Z}}, \frac{\partial (\mu_Z/\sqrt{D_Z})}{\partial D_Z} = -\frac{\mu_Z}{2} D_Z^{-\frac{3}{2}} \\ \frac{\partial \mu_Z}{\partial \mu^T} = [\frac{\partial \mu_Z}{\partial \mu_1}, \frac{\partial \mu_Z}{\partial \mu_2}, \cdots, \frac{\partial \mu_Z}{\partial \mu_n}]^T \\ \frac{\partial \mu_Z}{\partial D^T} = [\frac{\partial \mu_Z}{\partial D_1}, \frac{\partial \mu_Z}{\partial D_2}, \cdots, \frac{\partial \mu_Z}{\partial D_n}]^T \\ \frac{\partial D_Z}{\partial \mu^T} = [\frac{\partial D_Z}{\partial \mu_1}, \frac{\partial D_Z}{\partial \mu_1}, \cdots, \frac{\partial D_Z}{\partial \mu_1}]^T \\ \frac{\partial D_Z}{\partial D^T} = [\frac{\partial D_Z}{\partial D_1}, \frac{\partial D_Z}{\partial D_1}, \cdots, \frac{\partial D_Z}{\partial D_1}]^T \end{cases} \quad (17)$$

In respect of the GRENN method and thermal-structure coupling, the flowchart of the blisk LCF life reliability analysis is drawn in Figure 2 and its basic procedure is described below.

Step 1: Build the finite element (FE) model of blisk in a workbench environment;

Step 2: Consider the means of the input random variables (i.e., gas temperature, rotation speed, material parameters and fatigue performance parameters) and set boundary conditions to conduct the blisk FE analysis under the interaction of heat load, centrifugal load and then gain the minimum fatigue point as the design point of the blisk reliability design.

Step 3: Extract small samples of the input random variables using the Latin hypercube sampling (LHS) method and perform FE analyses for each group of samples to gain the output responses (blisk LCF life) and extract the minimum values of the responses as a training sample set by combining the input samples.

Step 4: Training the GRENN model by computing the optimal smooth factors, radial basis function and connection weights with the cross validation method [26], through the normalization of training samples.

Step 5: Structure of the limit state function of blisk LCF life with the established GRENN model.

Step 6: Check the precision of the GRENN model. If unacceptable, return to *Step 4*; if acceptable, conduct *Step 7*.

Step 7: Calculate the reliability degree and sensitivity degree of the fatigue life and input variables, by conducting the reliability and sensitivity analyses of blisk LCF life with thermal-structure coupling, through a large number of samples extracted by the MC method.

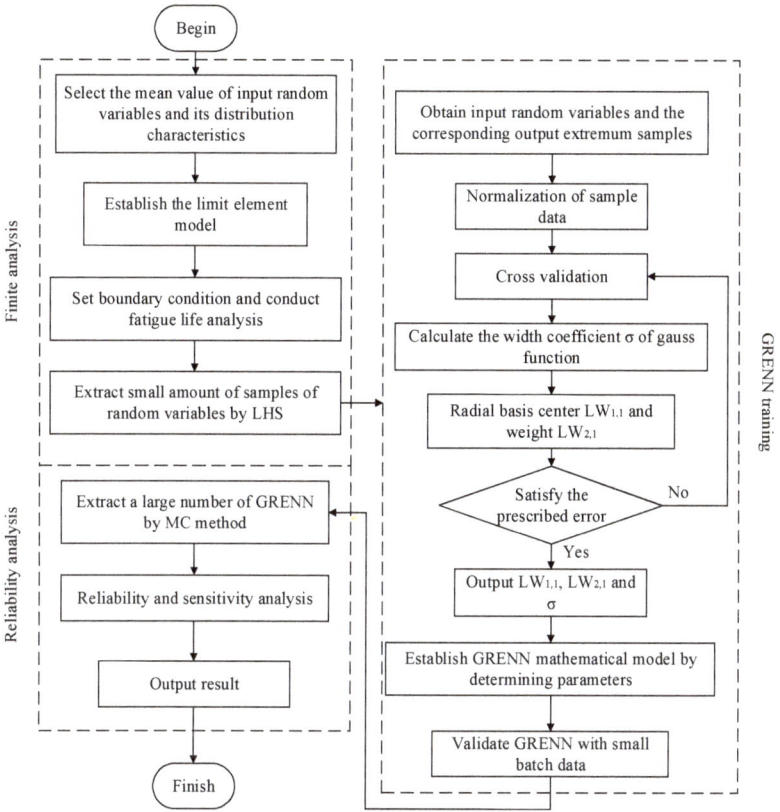

Figure 2. Flow chart of reliability analysis with GRENN method.

3. Reliability and Sensitivity Analyses of Blisk Low Cycle Fatigue Life

3.1. Random Variables Selection

In this study, we selected the high-pressure turbine blisk of an aeroengine with the high-temperature GH4133 as the object of study. In fact, the uncertainty and randomness of some parameters are the basic nature in blisk LCF life design and prediction [15]. By comprehensively regarding the engineering practice, the exiting data and the basic properties of parameter uncertainty studied by Repetto, et al. [15], the probabilistic analysis of the blisk LCF life was performed by the randomness of numerous reasonably-selected parameters, such as rotation speed ω, gas temperature T, material density ρ, heat conductivity λ, elasticity modulus E, fatigue strength efficient σ'_f, fatigue ductility coefficient ε'_f, fatigue strength index b and fatigue ductility index c. To simplify the calculation by combining engineering practices [43] and the present data, the selected variables were summed to be independent mutually and obey normal distributions. The distributions of the variables are listed in Table 1.

Table 1. Distributions of random variables.

Random Variables	Mean μ	Standard Deviation δ	Distribution
Density ρ, $kg \cdot m^{-3}$	8210	328.4	Normal
Rotate speed ω, $rad \cdot s^{-1}$	1168	35	Normal
Heat conductivity λ, $W \cdot m^{-1} \cdot {}^\circ C^{-1}$	23	0.005	Normal
Modulus of elasticity, E, MPa	163000	4890	Normal
Blade-root temperature T_a, k	1173.15	35.2	Normal
Blade-tip temperature T_b, k	1473.15	47	Normal
Fatigue strength efficient σ'_f	1419	42.5	Normal
Fatigue ductility coefficient ε'_f	50.5	1.53	Normal
Fatigue strength index b	−0.1	0.005	Normal
Fatigue ductility index c	−0.84	0.042	Normal

3.2. Deterministic Analysis of Blisk Low Cycle Fatigue Life

For the static analysis of the blisk, the blisk stress inducted by aerodynamic loads can be ignored because it is far less than that caused by the centrifugal load and heat load [43]. The deterministic analysis of the blisk was completed by regarding the interaction of temperature and centrifugal loads, under a workbench 16.0 environment in the computer with a central processing unit (CPU) mode of Xeon E5-2630V3 (Intel Corporation, Santa Clara, CA, USA) and RAM (Intel Corporation) of 64 GB. Due to the symmetry of the blisk, we selected 1/40 of the whole blisk for analysis to reduce the computational burden [44]. The FE models are shown in Figure 3, with 31,380 nodes and 17,111 elements. The thermodynamic analysis of the blisk was implemented in which the heat energy of a high temperature gas is transferred to the surface of the blisk according to the heat conduction law and heat convection. In light of thermodynamic theory, the temperature distribution on the blisk surface can be calculated by the empirical formula, i.e.,

$$T = T_a + (T_a - T_b)\left(\frac{R^m - R_a^m}{R_b^m - R_a^m}\right) \tag{18}$$

in which T_a is the temperature at blisk-root; R_a is the radius of blisk-root edge; T_b is the temperature at blisk-tip; R_b is the radius of blisk-tip; R is the radius of blisk in a different position; $m = 2$ was determined for the high temperature alloy GH4133B [39].

(a) FE model　　　　　　　　　　　(b) FE gridding

Figure 3. FE model and gridding of a turbine blisk.

By the displacement constraint of the blisk's inner diameter to restrict the degrees of freedom in the directions x, y and z, the deterministic analysis of the blisk was finished based on the means of the input variables in Table 1. The distributions of temperature, equivalent stress and equivalent strain are shown in Figure 4a–c. As seen in Figure 4a–c, the maximum stress of the blisk was 1 057.7 Mpa on the blade-root and the minimum strain was 8.142 7 × 10^{-3} m/m. Therefore, the node of the maximum strain on the blade-root was selected as the object of study for the blisk LCF life analysis. In terms of the Mason-Coffin formula in Equation (2) and the Miner line accumulative damage rule in Equation (3), the fatigue life values at the max-strain point of the blisk areshown in Figure 4d. It is illustrated in Figure 4d that the minimum fatigue life was 8900.6 cycles. In respect of the double safety coefficients in engineering, the LCF of the blisk should be about 4450 cycles based on the deterministic analysis.

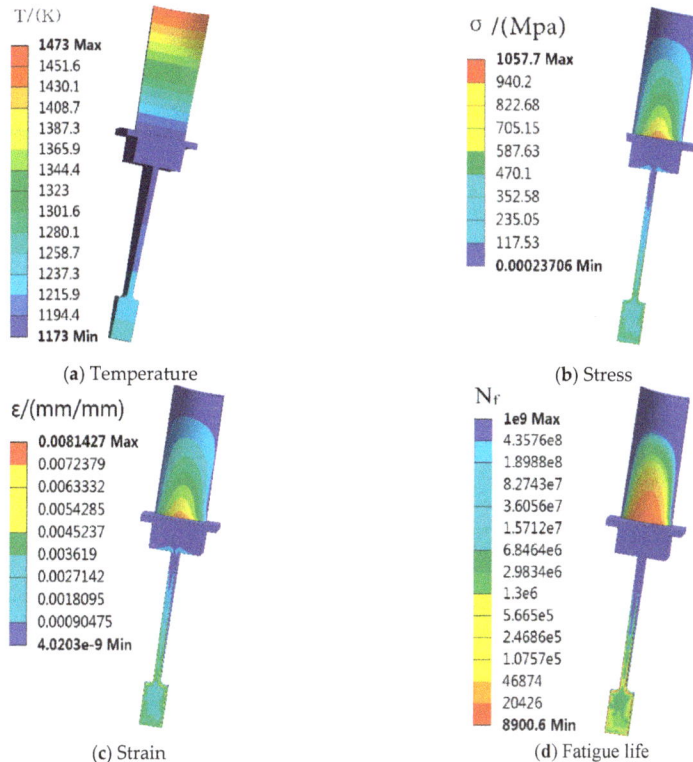

Figure 4. Nephgrams of the responses of blisk stress and fatigue life.

3.3. Low Cycle Fatigue Life Models of Blisk with GRENN Method

With regard to the distribution of the input random variables in Table 1, 150 samples (a small batch) were extracted by LHS technology. Based on these samples and FE analyses, the corresponding output responses (minimum LCF lives) were computed as the samples together with the extracted input samples. One hundred and twenty groups of samples were selected from the pool of training samples as training samples and the remain 30 groups of samples were selected as the test samples for the GRENN model.

Regarding the Gauss function as a transfer function in the hidden layer, the implicit layer weight $LW_{1,1}$ of the hidden layer was computed using the Euclidean distance method. The outputs of GRENN training were taken as the connection weights $LW_{2,1}$ between the hidden layer and the output layer. The original samples data should be normalized for each parameter. The normalized data were adopted to train the GRENN and then to gain the parameters of GRENN (the implicit layer weight $LW_{1,1}$ of the hidden layer, the connection weights $LW_{2,1}$ and the smooth factor σ) by the cross validation method [35], in which b and $LW_{1,1}$ and $LW_{1,2}$ (computed by Equations (6), (7) and (9)) are summarized in Equation(19). By inputting the values of these parameters into Equation (12), the GRENN model can be gained. The remaining 30 groups of samples were employed to test the established GRENN model. The prediction results are shown in Figure 5. From Figure 5, it can be seen that the predicted data were almost consistent with the true sample data, which indicates a high prediction precision for the developed GRENN model.

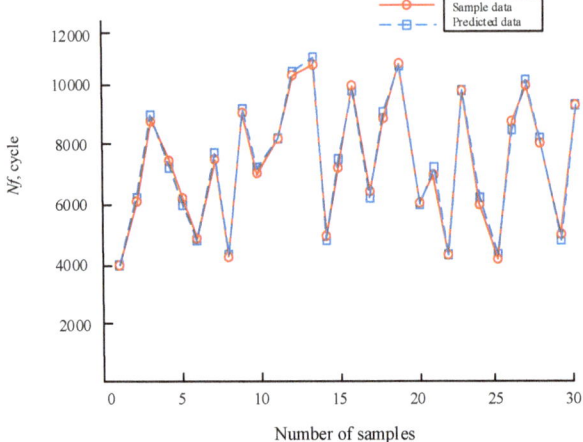

Figure 5. Predicted results of the GRENN model with 30 groups of samples.

$$\begin{cases} LW_{1,1} = \begin{bmatrix} -0.4189 & 0.0946 & \cdots & 0.9054 & 0.9459 & 0.7973 \\ -0.8912 & -0.5646 & \cdots & 0.7415 & -0.2517 & -0.1020 \\ 1.0000 & -0.5646 & \cdots & -0.7415 & 0.7959 & -0.5510 \\ -0.7852 & -0.7315 & \cdots & -0.8792 & -0.1678 & 0.4765 \\ 0.2245 & -0.8639 & \cdots & -0.2789 & -0.6190 & -0.3878 \\ 0.5839 & 0.3020 & \cdots & 0.2483 & 1.0000 & -0.7315 \\ 0.6510 & 0.2752 & \cdots & -0.6644 & 0.1678 & -0.5973 \\ -0.6892 & 0.8514 & \cdots & 0.6216 & -0.9865 & -0.4865 \\ -0.0470 & -0.5302 & \cdots & 0.7584 & 0.6107 & 0.0336 \\ -0.1757 & -0.8108 & \cdots & 0.7432 & 0.2838 & -0.3919 \end{bmatrix}^T_{10 \times 120} \\ LW_{2,1} = \begin{bmatrix} -0.9061 & -0.8701 & \cdots & -0.9627 & -0.9968 & -0.9400 \end{bmatrix}_{1 \times 120} \\ b = \begin{bmatrix} 2.8710 & 2.8710 & \cdots & 2.8710 & 2.8710 & 2.8710 \end{bmatrix}^T_{1 \times 120} \end{cases} \quad (19)$$

3.4. Reliability Analysis of Blisk Low Cycle Fatigue Life with GRENN Model

In this subsection, the reliability analysis of the blisk LCF life with the GRENN model was performed by 10,000 simulations with the MC method. The historical simulation diagram and histogram of the blisk LCF life are shown in Figure 6. As shown in Figure 6, the blisk minimum fatigue life followed a normal distribution with a mean of 9419 cycles and a standard deviation 967 cycles. As the allowable fatigue life $y^* = 6\,000$ cycles, the reliability degree P_r of the blisk LCF life was 0.99848 in line with Equations (13) and (15). The gained reliability degree basically catered for blisk design in engineering. In this case, the obtained fatigue life of a blisk was 6000 cycles in respect to the reliability analysis. However, the minimum LCF life of the deterministic analysis was 8900.6 cycles, as shown Figure 4d. In respect to the double safe coefficients, the safe fatigue life of blisk design was about 4450 cycles in engineering in line with the deterministic analysis. Therefore, it is revealed that the deterministic analysis method is backward-looking relative to ~4450 cycles of the probabilistic analysis method for blisk LCF life prediction at 6000 cycles, because 4450 cycles was far less than 6000 cycles.

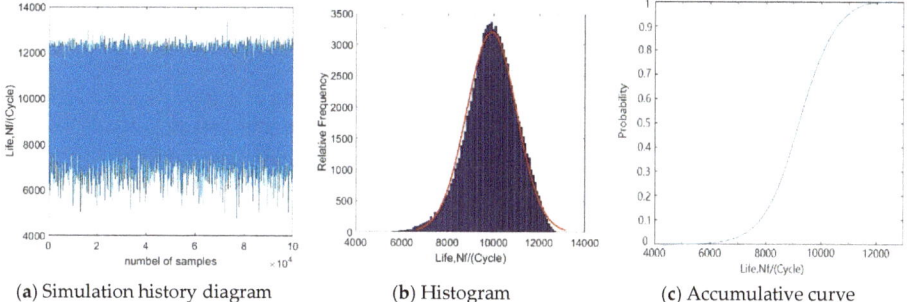

(a) Simulation history diagram (b) Histogram (c) Accumulative curve

Figure 6. Reliability analysis results of blisk fatigue life with the GRENN method.

3.5. Sensitivity Analysis of Blisk Low Cycle Fatigue Life with GRENN Method

Sensitivity reflects the level of sensitivity of the input random variables on blisk reliability, which is helpful to find the major impact factors and then guide structural design. Sensitivity involves the sensitivity degree and the effect probability. The sensitivity degree is defined by the effect of the input parameters on the output response with positive and negative signs. A positive sign indicates the input parameter was positively correlated with the output response and vice versa for a negative sign. Effect probability is defined as the ratio of the sensitivity degree of one input parameter to the total sensitivity degree of all input parameters. In terms of Equations (12)–(17), the sensitivity results are listed in Table 2 and Figure 7.

Table 2. Sensitivity degree and impact probability of the random input parameters.

Random Parameters	Sensitivity Degree, ×10^{-3}	Effect Probability, %
ρ	−0.41586	6.18
ω	−0.52565	7.81
λ	+0.0132	0.20
E	+0.16948	2.52
T	−1.76022	26.16
σ'_f	+0.41615	6.18
ε'_f	+0.21311	3.17
b	+0.43585	6.48
c	+2.7929	41.30

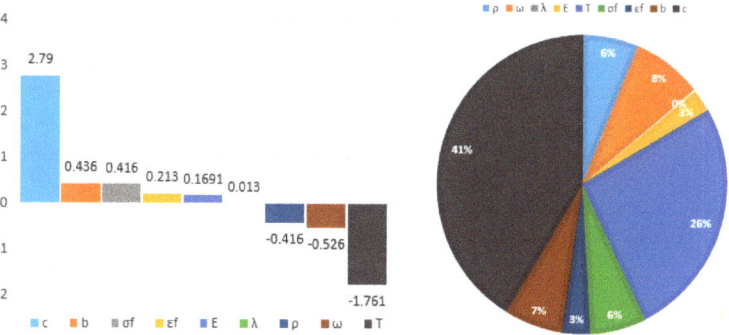

Figure 7. Sensitivity degree distributions of random parameters on blisk LCF life.

As demonstrated in Table 2 and Figure 7, the fatigue ductility index c and gas temperature T were two major influencing parameters because their effect probabilities and sensitivity degree were 41.3% and 0.27929, as well as 26.16% and −0.176122, respectively. Other parameters play small effects on blisk reliability. Therefore, T and c should be considered and controlled in blisk design with priority.

3.6. Validation of GRENN

To validate the effectiveness and validation of the GRENN method, the MC method and ERSM were used in the reliability analyses of blisk LCF life under different simulations based on the same computation conditions and random variables. The computing time and reliability degrees are presented in Tables 3 and 4. In Tables 3 and 4, the precision for the method D_p was computed under 10,000 simulations, by

$$D_p = 1 - \frac{|\gamma_a - \gamma_m|}{\gamma_a} \times 100\% \qquad (20)$$

in which γ_a is the reliability degree of the MC method; γ_m indicates the reliability degree of ERSM or the GRENN method.

Table 3. Computing time of the MC method, ERSM and GRENN.

Number of Samples	Computing Time under Different Simulations, s			Reduced Time, s	Improved Efficiency, %
	MC Method	ERSM	GRENN		
10^2	5400	1.249	1.201	0.048	3.843
10^3	14400	1.266	1.201	0.065	5.134
10^4	432000	1.681	1.311	0.370	15.18
10^5	—	2.437	1.342	1.095	44.93
10^6	—	4.312	2.138	2.174	50.42

Table 4. Computational precision of the reliability analysis methods under different simulations.

Samples	Reliability Degree			Precision/%		Improved Precision/%
	MC Method	ERSM	GRENN	ERSM	GRENN	
10^2	0.85	0.76	0.79	76.24	79.25	3.01
10^3	0.976	0.947	0.968	95.00	97.11	2.11
10^4	0.9968	0.9824	0.9973	98.56	99.95	1.39
10^5	—	0.98181	0.99848	98.49	99.83	1.34
10^6	—	0.98262	0.99587	98.58	99.91	1.33

As shown in Table 3, with the increasing number of simulations, the computing time increases for the MC method, ERSM and the GRENN method. For the MC method, the simulations larger than 10,000 require an excessive computational burden so that the MC method is unworkable for such large simulations. However, ERSM and the GRENN method only take a few seconds and thus can breezily implement simulations from 100 to 1,000,000. Relative to ERSM, the developed GRENN method spends less time and is highly computationally efficient, and that the strength of the GRENN method becomes more obvious with an increase in the number of simulations. For instance, under 10,000 simulations, the GRENN method reduces the computing time by 0.048 s and improves the computational efficiency by 3.843% relative to the ERSM, while the simulation time is reduced by 2.138 s and the efficiency is improved by 50.42%. Therefore, it is revealed that the proposed GRENN method has a strong computing power and is highly computationally efficient in probabilistic simulations. Meanwhile, the potential of high efficiency becomes stronger and more simulations are required.

In Table 4, the reliability degree computed by the MC method under 10,000 simulations is regarded as a reference. In this case, we find that the reliability degrees of ERSM and the GRENN method were 0.9824 and 0.9923 under 10,000 simulations, and their computational precisions were 98.56% and 99.95% so that the GRENN method improves the precision by 1.39%. Additionally, with an increasing number of simulations, the reliability degree of a turbine blisk increases and the developed GRENN method is more accurate than ERSM.

Therefore, the developed GRENN method is highly computationally precise and efficient and the strengths become more obvious for more simulations.

4. Conclusions

The aim of this paper was to propose a new reliability analysis method, i.e., the generalized regression extreme neural network (GRENN) method, for the reliability analysis of blisk LCF life, to improve the life and performance of turbine blisks. The developed GRENN absorbed the strengths of a generalized regression neural network (GRNN) in nonlinear mapping and small sample-based modeling, and the extremum response surface method (ERSM) for handling the transient problem of the dynamic reliability analysis of blisk LCF life. Through this study, some conclusions are summarized as follows:

(1) The reliability degree of blisk LCF life was 0.99848 when the life allowable value was 6000 cycles. Relative to 4450 cycles acquired from the deterministic analysis after considering the double coefficient of a safe life, the LCF (6000 cycles to ensure a reliability degree of 0.99848) of the blisk obtained from the reliability design had enough life margin (about 1550 cycles) to ensure the operation of the blisk structure.

(2) From the sensitivity analysis of a blisk, the fatigue ductility index c and gas temperature T played key roles in blisk LCF life evaluation and design. T and c were positively and negatively correlated with blisk life, respectively. The conclusions can significantly guide the optimization and design of blisk LCF life.

(3) Through the comparison of the methods, it is demonstrated that the developed GRENN method is far better than ERSM in modeling precision and computing efficiency and is basically consistent with the MC method. Moreover, the strengths of the GRENN method become more obvious with the increasing number of simulations. It is fully supported that the proposed GRENN method is a high-accuracy and high-efficiency method to address the key questions of nonlinearity, transients and large sample-based modeling.

In summary, the efforts of this paper provide a promising method (GRENN method) for the nonlinear dynamic reliability analysis of complex structures and enrich and develop mechanical reliability theory.

Author Contributions: Conceptualization, C.Z. and C.F.; methodology, C.Z.; software, J.W.; validation, J.W., H.J. and C.F.; formal analysis, J.W.; investigation, H.J.; resources, C.Z.; data curation, C.F.; writing—original draft preparation, C.Z. and W.T.; writing—review and editing, C.F.; visualization, H.J.; supervision, C.F. and W.T.; project administration, C.Z.; funding acquisition, C.Z. and C.F.

Funding: This research was funded by National Natural Science Foundation of China, grant numbers 51275138, 51605016 and 51475025, Start-up Research Funding of Fudan University, grant number FDU38341, Shanghai Pujiang Program, grant number PJ20191334 and The APC was funded by 51605016.

Conflicts of Interest: The authors declare that there are no conflict of interest regarding the publication of this article.

References

1. Bin, B.; Bai, G.C. Dynamic probabilistic analysis of stress and deformation for bladed disk assemblies of aeroengine. *J. Cent. South Univ.* **2014**, *21*, 3722–3735.
2. Gao, H.F.; Wang, A.; Bai, G.C.; Wei, C.M.; Fei, C.W. Substructure-based distributed collaborative probabilistic analysis method for low-cycle fatigue damage assessment of turbine blade-disk. *Aerosp. Sci. Technol.* **2018**, *79*, 636–646. [CrossRef]
3. Scott-Emuakpor, O.; George, T.; Cross, C.; Shen, M. Multi-axial fatigue-life prediction via a strain-energy method. *AIAA J.* **2010**, *48*, 63–72. [CrossRef]

4. Hou, N.X.; Wen, Z.X.; Yu, Q.M.; Yue, Z.Y. Application of a combined high and low cycle fatigue life model on life prediction of SC blade. *Int. J. Fatigue* **2009**, *31*, 616–619. [CrossRef]
5. Zhu, S.P.; Liu, Q.; Peng, W.; Zhang, X.C. Computational-experimental approaches for fatigue reliability assessment of turbine bladed disks. *Int. J. Mech. Sci.* **2018**, *142–143*, 502–517. [CrossRef]
6. Zhu, S.P.; Liu, Q.; Zhou, J.; Yu, Z.Y. Fatigue reliability assessment of turbine discs under multi-source uncertainties. *Fatigue Fract. Eng. Mater. Struct.* **2018**, *41*, 1291–1305. [CrossRef]
7. Zhu, S.P.; Liu, Q.; Lei, Q.; Wang, Q.Y. Probabilistic fatigue life prediction and reliability assessment of a high pressure turbine disc considering load variations. *Int. J. Damage Mech.* **2018**, *27*, 1569–1588. [CrossRef]
8. Zhu, S.P.; Huang, H.Z.; Smith, R.; Ontiveros, V.; He, L.P.; Modarres, M. Bayesian framework for probabilistic low cycle fatigue life prediction and uncertainty modeling of aircraft turbine disk alloys. *Probab. Eng. Mech.* **2013**, *34*, 114–122. [CrossRef]
9. Sun, Y.; Hu, L.S. Low cycle fatigue life prediction of a 300MW steam turbine rotor using a new nonlinear accumulation approach. In Proceedings of the 24th Chinese Control and Decision Conference (CCDC), Taiyuan, China, 23–25 May 2012; pp. 3913–3918.
10. Letcher, T.; Shen, M.H.H.; Scott-Emuakpor, O.; George, T.; Cross, C. An energy-based critical fatigue life prediction method for AL6061-T6. *Fatigue Fract. Eng. Mater. Struct.* **2012**, *35*, 861–870. [CrossRef]
11. Bargmann, H.; Rüstenberg, I.; Devlukia, J. Reliability of metal components in fatigue: A simple algorithm for the exact solution. *Fatigue Fract. Eng. Mater. Struct.* **2010**, *17*, 1445–1457. [CrossRef]
12. Zhu, S.P.; Foletti, S.; Beretta, S. Probabilistic framework for multiaxial LCF assessment under material variability. *Int. J. Fatigue* **2017**, *103*, 371–385. [CrossRef]
13. Zhu, S.P.; Huang, H.Z.; Peng, W.; Wang, H.K.; Mahadevan, S. Probabilistic Physics of Failure-based framework for fatigue life prediction of aircraft gas turbine discs under uncertainty. *Reliab. Eng. Syst. Saf.* **2016**, *146*, 1–12. [CrossRef]
14. Viadero, F.; Bueno, J.I.; de Lacalle, L.L.; Sancibrian, R. Reliability computation on stiffened bending plates. *Adv. Eng. Softw.* **1995**, *20*, 43–48. [CrossRef]
15. Pagnini, L.; Repetto, M.P. The role of parameter uncertainties in the damage prediction of the alongwind-induced fatigue. *J. Wind Eng. Ind. Aerodyn.* **2012**, *104–106*, 227–238. [CrossRef]
16. Repetto, M.P.; Torrielli, A. Long term simulation of wind-induced fatigue loadings. *Eng. Struct.* **2017**, *132*, 551–561. [CrossRef]
17. Marseguerra, M.; Zio, E. The cell-to-boundary method in Monte Carlo-based dynamic PSA. *Reliab. Eng. Syst. Saf.* **1995**, *48*, 199–204. [CrossRef]
18. Melchers, R.E.; Ahammed, M. A fast-approximate method for parameter sensitivity estimation in Monte Carlo structural reliability. *Comput. Struct.* **2004**, *82*, 55–61. [CrossRef]
19. Puatatsananon, W.; Saouma, V.E. Reliability analysis in fracture mechanics using the first-order reliability method and Monte Carlo simulation. *Fatigue Fract. Eng. Mater. Struct.* **2010**, *29*, 959–975. [CrossRef]
20. Pan, Q.; Dias, D. An efficient reliability method combining adaptive support vector machine and Monte Carlo simulation. *Struct. Saf.* **2017**, *67*, 85–95. [CrossRef]
21. Fei, C.W.; Bai, G.C.; Tian, C. Extremum response surface method for casing radial deformation probabilistic analysis. *AIAA J. Aerosp. Inf. Syst.* **2013**, *10*, 47–52.
22. Tvedt, L. Distribution of quadratic forms in normal space—Application to structural reliability. *J. Eng. Mech.* **1990**, *116*, 1183–1197. [CrossRef]
23. Fei, C.W.; Bai, G.C. Distributed collaborative extremum response surface method for mechanical dynamic assembly reliability analysis. *J. Cent. South Univ.* **2013**, *20*, 2414–2422. [CrossRef]
24. Kaymaz, I.; Marti, K. Reliability-based design optimization for elastoplastic mechanical structures. *Comput. Struct.* **2007**, *85*, 615–625. [CrossRef]
25. Fei, C.W.; Tang, W.Z.; Bai, G.C.; Shuang, M. Dynamic probabilistic design for blade deformation with SVM-ERSM. *Aircr. Eng. Aerosp. Technol.* **2015**, *87*, 312–321. [CrossRef]
26. Wei, Z.; Feng, F.; Wei, W. Non-linear partial least squares response surface method for structural reliability analysis. *Reliab. Eng. Syst. Saf.* **2017**, *161*, 69–77. [CrossRef]
27. Fei, C.W.; Bai, G.C. Distributed collaborative probabilistic design for turbine blade-tip radial running clearance using support vector machine of regression. *Mech. Syst. Signal Process.* **2014**, *49*, 196–208. [CrossRef]

28. Fei, C.W.; Choy, Y.S.; Hu, D.Y.; Bai, G.C.; Tang, W.Z. Dynamic probabilistic design approach of high-pressure turbine blade-tip radial running clearance. *Nonlinear Dyn.* **2016**, *86*, 205–223. [CrossRef]
29. Goswami, S.; Ghosh, S.; Chakraborty, S. Reliability analysis of structures by iterative improved response surface method. *Struct. Saf.* **2016**, *60*, 56–66. [CrossRef]
30. Bai, G.C.; Fei, C.W. Distributed collaborative response surface method for mechanical dynamic assembly reliability design. *Chin. J. Mech. Eng.* **2013**, *26*, 1160–1168. [CrossRef]
31. Hurtado, J.E.; Alvarez, D.A. An optimization method for learning statistical classifiers in structural reliability. *Probab. Eng. Mech.* **2010**, *25*, 26–34. [CrossRef]
32. Zhang, C.Y.; Bai, G.C. Extremum response surface method of reliability analysis on two-link flexible robot manipulator. *J. Cent. South Univ.* **2012**, *19*, 101–107. [CrossRef]
33. Lu, C.; Feng, Y.W.; Liem, R.P.; Fei, C.W. Improved kriging with extremum response surface method for structural dynamic reliability and sensitivity analyses. *Aerosp. Sci. Technol.* **2018**, *76*, 164–175. [CrossRef]
34. Nose-Filho, K.; Lotufo, A.D.P.; Minussi, C.R. Short-term multinodal load forecasting using a modified general regression neural network. *IEEE Trans. Power Deliv.* **2011**, *26*, 2862–2869. [CrossRef]
35. Zhao, C.; Liu, K.; Li, D.S. Freight volume forecast based on GRNN. *J. China Railw. Soc.* **2004**, *26*, 12–15.
36. Li, H.Z.; Guo, S.; Li, C.J.; Sun, J.Q. A hybrid annual power load forecasting model based on generalized regression neural network with fruit fly optimization algorithm. *Knowl. Based Syst.* **2013**, *37*, 378–387. [CrossRef]
37. Sun, G.; Hoff, S.J.; Zelle, B.C.; Smith, M.A. Development and comparison of backpropagation and generalized regression neural network models to predict diurnal and seasonal gas and PM 10 concentrations and emissions from swine buildings. *Trans. Asabe* **2008**, *51*, 685–694. [CrossRef]
38. Wang, Y.; Peng, H. Underwater acoustic source localization using generalized regression neural network. *J. Acoust. Soc. Am.* **2018**, *143*, 2321–2331. [CrossRef]
39. Gao, H.; Fei, C.; Bai, G.; Ding, L. Reliability-based low-cycle fatigue damage analysis for turbine blade with thermo-structural interaction. *Aerosp. Sci. Technol.* **2016**, *49*, 289–300. [CrossRef]
40. Liu, C.L.; Lu, Z.Z.; Xu, Y.L.; Yue, Z.F. Reliability analysis for low cycle fatigue life of the aeronautical engine turbine disc structure under random environment. *Mater. Sci. Eng. A* **2005**, *395*, 218–225. [CrossRef]
41. Vubac, N.; Lahmer, T.; Keitel, H.; Zhao, J.; Zhuang, X.; Rabczuk, T. Stochastic predictions of bulk properties of amorphous polyethylene based on molecular dynamics simulations. *Mech. Mater.* **2014**, *68*, 70–84. [CrossRef]
42. Zhai, X.; Fei, C.W.; Zhai, Q.G.; Bai, G.C.; Ding, L. Reliability and sensitivity analyses of HPT blade-tip radial running clearance using multiply response surface model. *J. Cent. South Univ.* **2014**, *21*, 4368–4377. [CrossRef]
43. Shu, M.A.; Wang, K.M.; Hui, M.; Shuai, Z. Research on turbine blade vibration characteristic under steady state temperature field. *J. Shenyang Aerosp. Univ.* **2011**, *28*, 18–21.
44. Zhang, C.Y.; Lu, C.; Fei, C.W.; Liu, L.J. Multiobject reliability analysis of turbine blisk with multidiscipline under multiphysical field interaction. *Adv. Mater. Sci. Eng.* **2015**, *2015*, 519–520. [CrossRef]

© 2019 by the authors. Licensee MDPI, Basel, Switzerland. This article is an open access article distributed under the terms and conditions of the Creative Commons Attribution (CC BY) license (http://creativecommons.org/licenses/by/4.0/).

Article

Application of a New, Energy-Based ΔS* Crack Driving Force for Fatigue Crack Growth Rate Description

Grzegorz Lesiuk

Faculty of Mechanical Engineering, Department of Mechanics, Materials Science and Engineering, Wroclaw University of Science and Technology, PL-50370 Wrocław, Poland; Grzegorz.Lesiuk@pwr.edu.pl; Tel.: +48-71-320-3919

Received: 31 December 2018; Accepted: 7 February 2019; Published: 9 February 2019

Abstract: This paper presents the problem of the description of fatigue cracking development in metallic constructional materials. Fatigue crack growth models (mostly empirical) are usually constructed using a stress intensity factor ΔK in linear-elastic fracture mechanics. Contrary to the kinetic fatigue fracture diagrams (KFFDs) based on stress intensity factor K, new energy KFFDs show no sensitivity to mean stress effect expressed by the stress ratio R. However, in the literature there is a lack of analytical description and interpretation of this parameter in order to promote this approach in engineering practice. Therefore, based on a dimensional analysis approach, ΔH is replaced by elastic-plastic fracture mechanics parameter—the ΔJ-integral range. In this case, the invariance from stress is not clear. Hence, the main goal of this paper is the application of the new averaged (geometrically) strain energy density parameter ΔS^* based on the relationship of the maximal value of J integral and its range ΔJ. The usefulness and invariance of this parameter have been confirmed for three different metallic materials, 10HNAP, 18G2A, and 19th century puddle iron from the Eiffel bridge.

Keywords: fatigue crack growth; mean stress effect; J-integral; energy approach; generalized Paris' Law; crack growth rate; R-ratio

---

1. Introduction

Fatigue and fatigue cracking are the two main (more than 70% of all failures) phenomena responsible for the process of destroying steel structures. The fatigue crack growth phase is an essential process for the long-term operating structures subjected to cyclic loading. A fatigue crack may be formed either as a result of the accumulation of fatigue damage (intrusions and extrusions) or as a result of manufacturing processes. The appearance of fatigue cracks can also be the result of unfavorable operating conditions. The mere fact that a crack exists does not necessarily (anymore) mean that such an element (still referred to as metallic construction materials) has to be eliminated from service. The period of precritical fatigue crack growth can be expressed in a general way using an integral:

$$N_{cr} = \int_{a_o}^{a_{cr}} \frac{da}{f(\sigma_{ext}, a, P_{fc}, Y, R)} \quad (1)$$

where: N_{cr} is precritical fatigue crack growth period, a_o is the crack length, a_{cr} is the critical crack length, σ_{ext} is the external load, P_{fc} is the fracture mechanics parameter, the so-called crack driving force (CDF), Y is the geometric constraint, and R ($\sigma_{min}/\sigma_{max}$) is the stress ratio.

In the case of static loads, it is crucial to determine the critical load that triggers the avalanche development of the crack or to look for the critical length of the crack at which the element will continue to carry the assumed load. For safety reasons, an important issue is to determine the subcritical time of fatigue crack development under cyclic loading condition. According to the above, it is necessary to solve Equation (1) with known boundary conditions. However, in this case, it is crucial to provide a fatigue crack growth law that is as robust as possible. In the 1960s, Paris [1] correlated a quantity derived from classical fracture mechanics—ΔK stress intensity factor range (SIF)—with the fatigue crack growth rate from experimental data. Paris proposed this relationship in the form of a power-law function—known in the literature as the Paris law:

$$\frac{da}{dN} = C(\Delta K)^m \qquad (2)$$

Engineers are predicting the life of an element with a defect using various analytical and numerical techniques (including FEM and BEM). In Equation (2) C, m are constants determined from the Kinetic Fatigue Fracture Diagram (KFFD) presented in Figure 1. ΔK is related to the range of load changes $\Delta K = K_{max} - K_{min}$, corresponding successively to the range of external load changes.

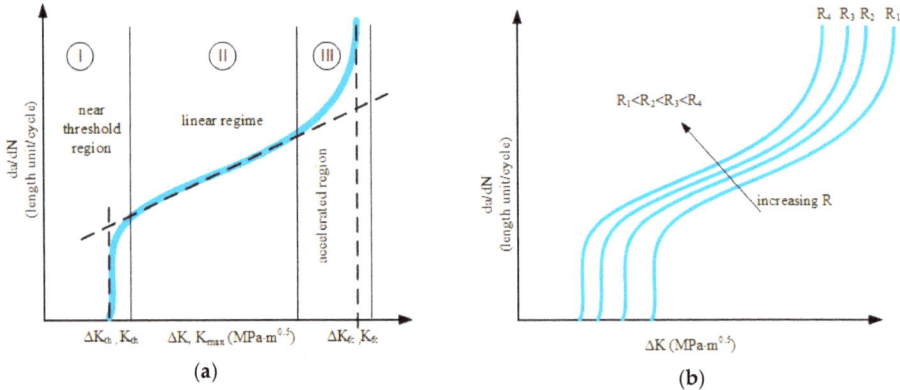

Figure 1. Schematic Kinetic Fatigue Fracture Diagram (KFFD): (a) sigmoidal shape of the fatigue crack growth rate (FCGR) curve; (b) the scheme of the stress R-ratio impact on the FCGR curves ($R_4 > R_3 > R_2 > R_1$).

The constant m in the Paris' law model is related to the angle of inclination of the experimental data straight line (Figure 1a), while the constant C is the ordinate at the intersection of the extension straight line from Area II. Generally speaking, the typical graph of fatigue cracking kinetics is divided into three areas (Figure 1a). Area I is the domain of the so-called low crack growth rate, i.e., from 0 to approximately 10^{-9} m/cycle, while Area II corresponds to the straight line of average fatigue crack growth rate in the range 10^{-9}–10^{-6} m/cycle—so called Paris regime. Area III is the domain of high fatigue crack growth rate, i.e., above 10^{-5} m/cycle. These ranges are contractually accepted as they depend on the material, its properties, the environment, and the load itself. The range (I) is limited on the left by the asymptote corresponding to the threshold ΔK_{th} value—the threshold range of stress intensity coefficient below which fracture does not propagate or propagates at an insignificantly low speed. The development of cracking ends when the stress factor reaches the critical value ΔK_{fc}, above which cracking propagates unstable. The effect of the stress ratio R factor is schematically represented as in Figure 1b. For higher R-factors, the ΔK_{th} threshold value, which triggers the fatigue cracking process, is lower, and all da/dN-ΔK curves are shifted. Therefore, the adaptation of Equation (2), taking into account the effect of the cycle asymmetry coefficient, is a major topic in Fatigue Fracture Mechanics (FFM). One of the well known is the solution proposed by Forman [2]:

$$\frac{da}{dN} = \frac{C(\Delta K)^m}{(1-R)K_C - \Delta K} \qquad (3)$$

The K_c quantity in Equation (3) is the fatigue fracture toughness (or critical stress intensity factor) for the given load conditions. In the event of difficulties in its establishment, it is very often replaced by known K_{IC}. Accurate determination of constants m and C requires knowledge of crack speed courses for different cycle asymmetry coefficients. Another proposal is well-known as Walker's law [3]:

$$\frac{da}{dN} = \frac{C_w}{(1-R)^{n_w}} (\Delta K)^{m_r} \qquad (4)$$

The constant C_w is determined here experimentally for different values of R. For $R = 0$ this constant is equivalent to the Paris' constant C. The exponent m_r is treated as a constant, similarly n_w—is obtained from experimental data. It is also determined by the extrapolation of data from kinetic fatigue fracture diagrams constructed for different R-ratios.

The influence of the cycle asymmetry coefficient strongly determines the analytical description of the fatigue gap closing process noticed by Wolf Elber [4]. The model covering the above issues is the model known in the literature on the subject of Forman-Mettu [5]:

$$\frac{da}{dN} = C \left[\left(\frac{1-f}{1-R} \right) \Delta K \right]^n \frac{\left(1 - \frac{\Delta K_{th}}{\Delta K}\right)^p}{\left(1 - \frac{K_{max}}{K_c}\right)^q} \qquad (5)$$

In the Forman-Mettu model expressed in Equation (5), c, n, p, and q are experimentally determined constants, and f is associated with the function of crack opening. The form of this function can be determined based on, e.g., FITNET procedures [6]. However, the application of this model is not easy—mainly due to a large number of experimentally determined constants.

A different group of attractive fatigue fracture models are the mathematical models based on the relationship of low cycle fatigue (LCF) parameters with the fatigue crack propagation rate. Fatigue crack growth is considered an elementary act in the local plastic zone of fracture including two different areas: static, corresponding to the maximum value of load in the cycle, and cyclic, corresponding to the amplitude of load. This concept is presented in Figure 2. Several authors [7–14] proposed excellent relationships between LFC (Low Cycle Fatigue) and FCGR (Fatigue Crack Growth Rate) data. However, in all cases, the presented models work well for the maintained R-ratio. It is more likely that the main reason of the mathematical collapse in R-ratio invariance of the fatigue crack growth description is associated with the fact that the crack driving force still depends only on the ΔK range or on the maximum value of K.

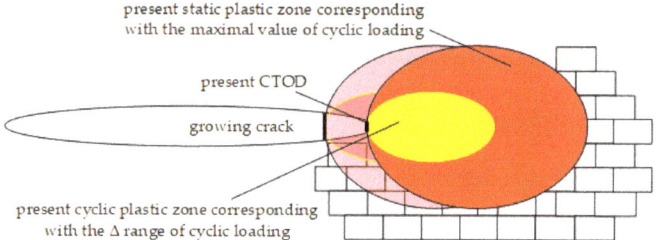

Figure 2. Schematic representation of plastic zones ahead of a fatigue crack tip with the marked Crack Tip Opening Displacement CTOD—δ.

Kujawski [15,16] proposed a new, crack driving force for a ΔK description of FCGR—the geometrical mean value of positive ΔK and K_{max}. The next development of the proposed model was the introduction of the weighting exponent α [17,18]:

$$K^* = \left(\Delta K^+\right)^{1-\alpha} \cdot K_{max}^\alpha \qquad (6)$$

According to the above, it is worth underlining an excellent contribution of Kujawski's model into the force approach combaning positive part of the stress intensity factor range ΔK-ΔK^+ and K_{max}. However, the meaning of the α parameter (ranged from 0 to 1) is debatable and in each case should be calibrated from da/dN-ΔK curves.

Therefore, the main aim of this work is to propose a new, crack driving force parameter with a good physical interpretation responsible for the fatigue fracture process without R-ratio influence.

Alternative methods of describing the kinetics of fatigue cracking are also being searched in order to eliminate the problem of avoiding the R-ratio effect. Research conducted by the author shows that it is possible to obtain such a dimensional base for KFFD, in which the description of fatigue cracking kinetics will not depend on the stress ratio. Energy as a dimensional quantity combining the dimensions of "force" and "displacement" seems to be naturally predestined to describe the kinetics of cracking. Many of the hypotheses concerning both fatigue and the description of fatigue cracking are based on energy irreversibly dissipated in each cycle of the load spectrum [19,20]. The dissipated energy accumulated in the material can be recorded as a hysteresis loop during the test. Determination of the subcritical period of fatigue crack growth requires the application of the first principle of thermodynamics. Assuming an infinite solid body model subjected to sinusoidal alternating external loads with a central part-thru, this balance can be formulated as follows [21–24]:

$$A + Q = W + K_e + \Gamma \tag{7}$$

In Equation (7), A represents the work of external loads after N cycles, Q represents the heat supplied to the body during loading, and W is the deformation energy after N cycles. The kinetic energy of the body is marked as K_e. Γ is the damage energy necessary for fatigue crack growth increment. When formulating the energy balance expressed in Equation (7), according to [23], the quantities A, Q, W, and K_e, are referred to unit of thickness. It is also assumed that the fatigue crack will grow slowly during each cycle so that the kinetic energy and heat exchanged during this process can be disregarded (i.e., for low loading frequencies). After the differentiation of Equation (7) and simplifications, the energy balance can be represented by

$$\frac{\partial A}{\partial N} = \frac{\partial W}{\partial N} + \frac{\partial \Gamma}{\partial N} \tag{8}$$

However, damage energy Γ can be expressed as a sum of static and cyclic components:

$$\Gamma = W_c + W_s \tag{9}$$

The energy of the static (monotonic) component of cyclic plastic deformations W_s, corresponding to the maximal value of external loading, is considered as the maximum value of the dissipation of the static energy activating the fracture process without the energy of cyclic deformation changes ($W_c = 0$) [24]. W_c corresponds to the dissipated energy during cyclic loading. For fatigue crack growth, both quantities are equally important. Hence, the final form of the fatigue crack growth surface can be expressed as

$$\frac{\partial S}{\partial N} = \frac{\partial W_c / \partial N}{\partial (\Gamma - W_s) / \partial S} \tag{10}$$

$$dS/dN = \frac{\alpha W_c^{(1)}}{\sigma_{plf} \varepsilon_{fc} (1 - K_{I\,max}^2 / K_{fc}^2)} \tag{11}$$

Thus, the proposed crack driving force ΔH (expressed in J/m^2) is equal:

$$\Delta H = \frac{W_c^{(1)}}{B(1 - K_{I\,max}^2 / K_{fc}^2)} \tag{12}$$

The Paris-like model can be represented by

$$da/dN = A(\Delta H)^k \tag{13}$$

In the above equations, α is a constant dimensionless factor, S is the fatigue crack area, σ_{plf} is the cyclic yield point, ε_{fc} is the critical strain value under cyclic conditions, and ΔH is the new energy parameter—the crack-driving force. In Equation (13), k should be equal to 1 (based on dimensional analysis approach). However, it is also reported in [22] that k varies from 0.87 to 1.38 for different ductility levels of the tested materials. On the other hand, in the original approach [24], this parameter strongly depends on the hysteresis loop area measured in the load line. It is more likely that this approach seems to be correct from a physical point of view but strongly depends on the experimental techniques of the registered dissipated energy. The problem of measuring dissipated energy regarding the global-local energy approach is widely discussed in [19]. There are no doubts that the energy approach supported by numerical methods is a powerful tool in a proper description of FCGR and residual lifetime estimation based on a rather physical not an empirical model. So far, no attempts have been made to analyze the ΔH characterization of the parameter, to link it with well-known parameters from classical fracture mechanics, i.e., CTOD or J-integral. This fact explains, among other reasons, the relatively low acceptance of energetic models in engineering practice. From the physical point of view, the other well interpretable quantity with this physical dimension ΔH is the integral J. Moreover, contemporary numerical and full field experimental methods allow determining the integral J for advanced materials and loading cases in an efficient way. Many times in the range of linear fracture mechanics, integral J allows one to determine stress intensity factors where there are no closed-form analytical solutions. This has been demonstrated in the author's and co-author's works [25–28] devoted to mixed mode fatigue crack growth description. However, it is worth noting that the number of kinetic energy models based on integral J is still negligible. Dowling and Begley [29] were the pioneers in describing the kinetics of fatigue cracking [29] using the cyclic J integral concept. At a later stage, Dowling [30] demonstrated the independence of integral J as a CDF (Crack Driving Force) from the geometry of samples, which encouraged the stability of this size in the description of the kinetics of fatigue curing. The path independence in applications to fatigue crack growth problems is also proven in many papers [31–34]. However, also in this case, the full invariability with respect to the R-ratio coefficient is not always obtained by substitution of the crack driving force from ΔK to the integral ΔJ. As evidence, the experimental data for 18G2A and 10HNAP steels (according to standard Polish nomenclature) recorded in papers [35–39] using bended specimens in FCGR experiment in a linear and nonlinear fracture mechanics validity range.

Independently from Rozumek's results [36–39], Joyce et al. [40] reported for cast stainless steel equivalent to ASME SA-351CF8M a similar impact of R-ratio in elastic-plastic cyclic ΔJ-crack driving obtained from standardised compact tension (CT) specimens. On the other hand, there are no doubts that the J-integral crack driving force can be applied where the validity of the linear elastic fracture mechanics is limited, so the ΔK approach is debatable. However, experimental observations [29–40] have confirmed the fact that the J-integral approach does not entirely solve the problem of the mean stress effect in the description of the FCGR. On the other hand, many analytical models seem to confirm the fact that the solution in the unification of the fatigue cracking process should be sought not so much in the amplitude of the force driving the crack but in the mutual combination of maximum CDF values and CDF amplitude.

Considering the above, the previously formulated basics of energy modeling allowed for the determination of the relation between fatigue fracture rate and energy parameters. Details of the dimensional analysis modeling [41,42] are available in [21–24,43,44]. In these works, it was demonstrated that the required energetic parameter with the physical dimension (J/m^2) is responsible for the fatigue fracture process. As it is based on the energy measured in the form of a local hysteresis loop, the ambiguities with multi-axis loads and geometries of the specimens did not allow for its proper application. On the other hand, the concept initiated by Kujawski [16] for the elastic parameter

ΔK seems to be worth considering when applying for elastic-plastic states. From experimental works and methods of energy accumulation [45], it can be indicated that schematically variable average load connects ΔJ and J_{max}. During fatigue crack growth, as is shown in Figure 3, under a different load ratio R, in the case of load paths from point i to j (negative maximal load value), the damage accumulation cannot cause crack growth (no crack opening means that ΔJ^+ is zero). For the load case from k to l (negative stress ratios), the crack is partially open (excluding the Elber crack closure phenomenon) in the positive part of the loading ($\Delta J^+ < \Delta J$). For the positive R-ratios (load path from m to n), the ΔJ^+ is equal to ΔJ. It is worth noting that, during a cyclic load, J_{max} and ΔJ are the values that bind the local stress intensity in front of the crack front—J plays the same role as K in the elastic-plastic fracture mechanics. Therefore, the geometric mean of ΔJ^+ and J_{max} is proposed as a new crack driving force candidate:

$$\Delta S* = \sqrt{\Delta J^+ \cdot J_{max}} \tag{14}$$

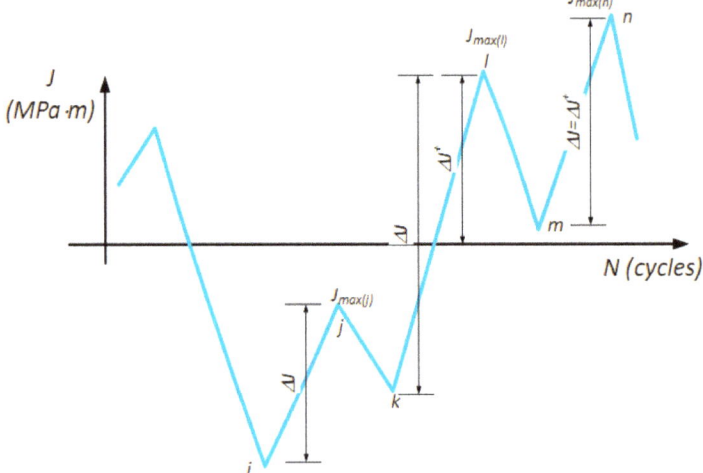

Figure 3. Schematic representation of the $\Delta S*$ components during cyclic loading.

2. Kinetic Fatigue Fracture Diagrams for 10HNAP, 18G2A, and the Eiffel bridge 19th-Century Puddle Iron

For experimental verification, the published fatigue crack growth data (for steel 10HNAP and 18G2A) based on the ΔJ integral range [35–38] was used. In Rozumek's papers [35–38], the test fatigue crack growth experiment based on elastic-plastic fracture mechanics was performed on the fatigue test stand MZGS-100 (Figure 4), enabling cyclically variable and static (mean) loading.

In an experimental campaign (based on ΔJ parameter as a crack driving force), a bent beam (Figure 5) made from 10HNAP and 18G2A steel was used. The specimens subjected to bending had an external unilateral sharp notch of 5 mm in depth, with the rounding radius $\rho = 0.5$ mm. The specimen dimensions were length $L = 120$ mm, width $W = 20$ mm, and thickness $t = 4$ mm. The scheme of the specimen is presented in Figure 5.

In the cited experimental campaign, the total ΔJ parameter [24,35,36] was considered as a sum of the elastic and plastic components of the cyclic ΔJ parameter [24]:

$$\Delta J = \Delta J_e + \Delta J_p = 2\pi Y^2 a \left(\int_0^{\Delta \varepsilon_e} \sigma d\varepsilon_e + \int_0^{\Delta \varepsilon_p} \sigma d\varepsilon_p \right) \tag{15}$$

In Equation (15) ΔJ_e represents elastic part of ΔJ integral range, ΔJ_p – plastic part, Y—geometric constraint, ε_e—elastic strains, ε_p—plastic strains measured in the vicinity of a crack tip. The tests were performed on the fatigue test stand MZGS-100 (Figure 3) with the loading frequency 29 Hz and the maximal bending moment was equal: M_a = 15.64 nm under the Mode I condition. Three different R-ratios were considered; $R = 0$, $R = -0.5$, $R = -1$. During the experiments, the crack length was observed using optical methods. All experimental details can be found in Rozumek's papers [36–39]. After experiments, the elastic-plastic J integral was calculated using boundary element methods (BEM) and FRANC3D software for fracture mechanics analysis. The detailed experimental-numerical procedure is described in [36–39]. Finally, the elastic-plastic, kinetic fatigue fracture diagrams were constructed based on the J-parameter.

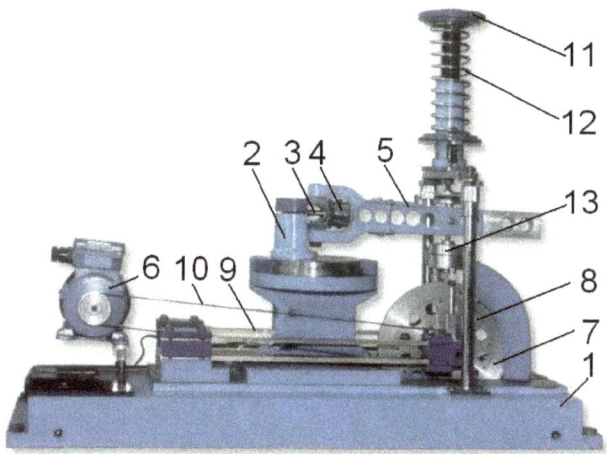

Figure 4. Fatigue testing machine MZGS-100: 1—bed, 2—rotational head with a holder, 3—specimen, 4—holder, 5—lever (effective length = 0.2 m), 6—motor, 7—rotating disk, 8—unbalanced mass, 9—flat springs, 10—driving belt, 11—spring actuator, 12—spring, 13—hydraulic connector [36].

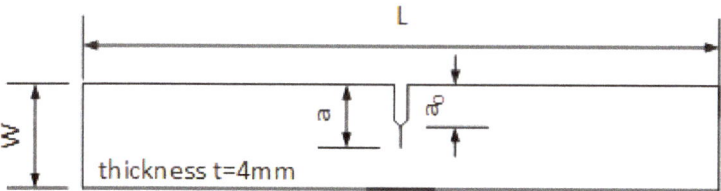

Figure 5. Scheme of the specimen subjected for the bending test in Gasiak & Rozumek [35] experiment, L = 120 mm, W = 20 mm, notch length a_0 = 5 mm (notch tip angle 60°, root radius ρ = 5 mm), a—fatigue crack length.

As an alternative, for the ΔS^* crack driving force, the kinetic fatigue fracture diagrams were parallel-constructed. The FCGR experiment for the Eiffel bridge [46] puddle iron was designed and performed in accordance with the American standard ASTM E647 [47]. Scheme of the specimen is presented in Figure 6. The stress intensity factor (SIF) for the standardized Compact Tension (CT) specimen is described by [47]

$$\Delta K = \frac{\Delta F}{B\sqrt{W}} f(a/W) \tag{16}$$

$$f\left(\frac{a}{W}\right) = \frac{(2+\alpha)(0.886 + 4.64\alpha - 13.32\alpha^2 + 14.72\alpha^3 - 5.6\alpha^4)}{\sqrt{(1-\alpha)^3}} \quad (17)$$

where α is the normalized crack length referred to the specimen width ($\alpha = a/W$), B is the thickness of the specimen, W represents the specimen width, and ΔF is the applied force range. The crack length was monitored using compliance methods from the clip-gage mounted on additional knives on the front side of the specimen—CMOD (crack mouth opening displacement).

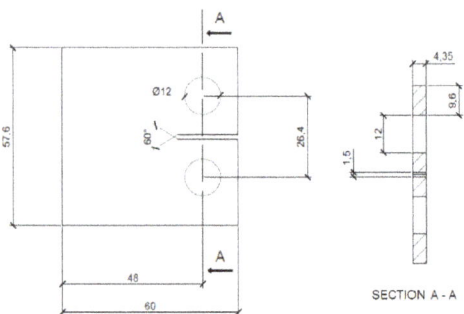

Figure 6. Compact Tension (CT) specimen scheme and dimensions for the Mode I FCGR experiment (puddle iron, all dimensions in mm).

For the energy fatigue crack growth description (puddle iron from the Eiffel bridge), only the elastic part of the ΔJ-integral range was analyzed using a well-known relationship from linear-elastic fracture mechanics (with assumed plane stress conditions):

$$\Delta J_e = \frac{\Delta K^2}{E} \quad (18)$$

The chemical composition of analyzed materials is presented in Table 1. Tables 2 and 3 include both static and cyclic mechanical properties of the considered materials. Kinetic fatigue fracture diagrams for all materials are shown in Figures 7–9.

Table 1. Chemical composition (in % by weight) of the tested materials.

Material	C	Mn	Si	P	S	Cr	Ni	Cu	Fe
10HNAP [37]	0.14	0.88	0.31	0.066	0.027	0.73	0.30	0.345	bal.
18G2A [36]	0.20	1.49	0.33	0.023	0.024	0.01	0.01	0.035	bal.
tested puddle iron	0.01	0.01	0.07	0.354	0.045	-	-	-	bal.

Table 2. Static mechanical properties of the analyzed metallic materials.

Material	Ultimate Tensile Strength UTS (MPa)	Yield Strength $\sigma_{pl}/\sigma_{0.2}$ (MPa)	Young Modulus E (GPa)	Poisson Ratio ν (-)	Fracture Toughness J_{IC} (MPa·m)
10HNAP [37]	566	418	215	0.29	0.178
18G2A [36]	535	357	210	0.3	0.320
Eiffel Bridge puddle iron [46]	342	292	193	0.3	n/a

Table 3. Low cycle fatigue properties of the analyzed materials.

Material	Fatigue Strength Coefficient σ_f' (MPa)	Fatigue Ductility Coefficient ε_f' (-)	Fatigue Strength Exponent b (-)	Fatigue Ductility Exponent c (-)
10HNAP [37]	746	0.442	−0.080	−0.601
18G2A [36]	782	0.693	−0.118	−0.410
Puddle Iron Eiffel Bridge [46]	603	0.160	−0.078	−0.797

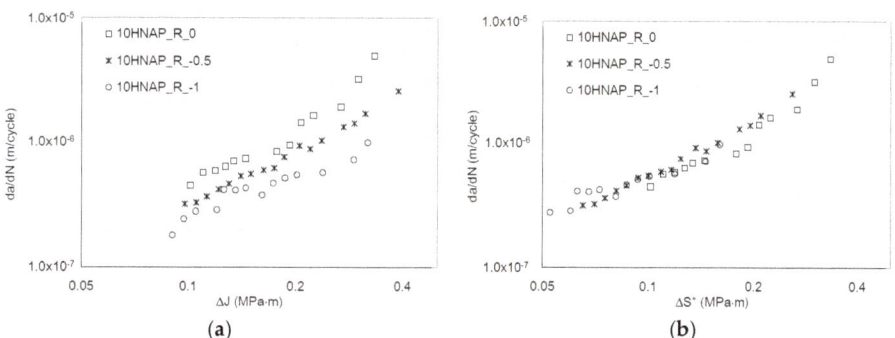

Figure 7. Fatigue crack growth curves for 10HNAP steel (**a**) based on the ΔJ crack driving force (based on data from [37]) and (**b**) based on the new, averaged ΔS^* crack driving force.

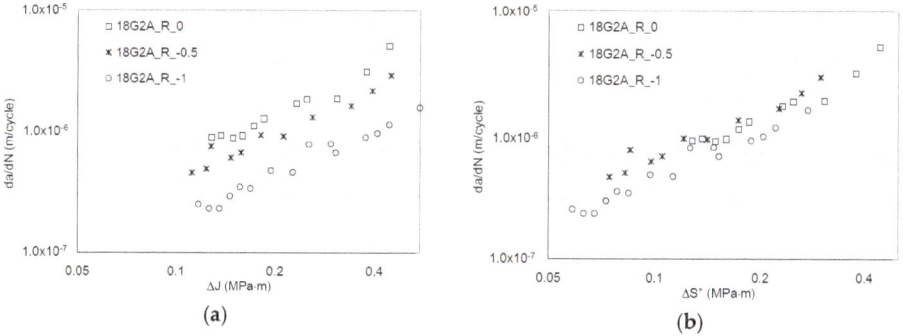

Figure 8. Fatigue crack growth curves for 18G2A steel (**a**) based on the ΔJ crack driving force (based on data from [37] and (**b**) based on the new, averaged ΔS^* crack driving force.

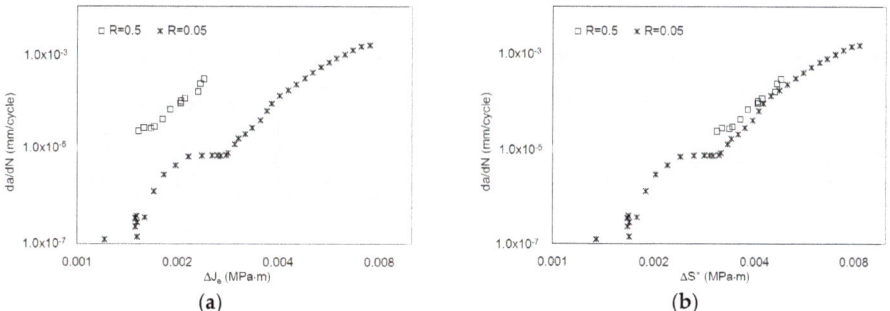

Figure 9. Fatigue crack growth curves for the puddle iron from the Eiffel bridge (**a**) based on the ΔJ crack driving force and (**b**) based on the new, averaged ΔS^* crack driving force.

As observed, the impact of R is noticeable in the case of the ΔJ description in the FCGR diagrams. In the same way, a similar tendency in the case of the da/dN-J_{max} description using the relationship between J_{max} and ΔJ can be demonstrated. However, in the case of an energy approach, the main concept assumed the representation of the fatigue crack growth rate using invariant kinetic fatigue fracture diagrams based on a new, crack driving force—ΔS^*. ΔS^* involves both ΔJ and J_{max} parameters. However, ΔJ should be limited only for ΔJ^+. Of course, possible corrections can be done using an effective ΔJ^+ range based on crack closure measurements. Recently, in [48,49], an efficient experimental method was described for in-situ J-integral calculation using DIC—digital image correlation. In the presented case, in order to analyze the invariance ability from stress ratio R, the crack driving force was examined by the simple statistical R^2 data fitting using a Paris-like relationship, for ΔJ and ΔS^* crack driving forces:

$$\frac{da}{dN} = A_1 (\Delta J)^n \qquad (19)$$

$$\frac{da}{dN} = D(\Delta S*)^p \qquad (20)$$

As was expected in each case, the new ΔS^* crack driving force consolidated all experimental results for different R-ratios into one straight line in the Paris regime. Moreover, the Paris-like law data fitting is better in the case of the description of the FCGR curves using the newly proposed crack driving force. Statistical outputs of the R-square data fitting to the Equations (19) and (20) are presented in Table 4. According to the statistical analysis, in each case, the introduction of the ΔS^* parameter caused a significant increase in data fitting for the Paris-like model expressed in Equation (20) in comparison to the model expressed in Equation (19).

Table 4. Statistical analysis of data fitting for the Kinetic Fatigue Fracture Diagram (KFFD) based on ΔJ and ΔS^* (all R-ratios).

All Data from FCGR Tests	R^2 da/dN-(ΔJ) Equation (19)	R^2 da/dN-(ΔS^*) Equation (20)
10HNAP	0.73	0.94
18G2A	0.54	0.91
Puddle Iron from the Eiffel Bridge	0.64	0.97

3. Conclusions

This article presents an understanding of dimensional analysis identifying the energy force driving the crack. In the course of this work, it was established that this value is the parameter of ΔH (J/m^2) proposed by Szata [24]. In the absence of an unambiguous analytical interpretation of this parameter, it was replaced by the integral J, which did not allow avoiding the influence of average load in the form of R. A new CDF—ΔS^*—based on the geometric mean of the maximum value of cyclic J-integral and its range ΔJ has been proposed to describe the kinetics of fatigue cracking. Its suitability for three different engineering materials (10HNAP and 18G2A steels and puddle iron) has been tested. On the basis of the above considerations, the following conclusions may be formulated:

- In contrast to J, S^* unambiguously describes the fatigue crack kinetics for 10HNAP and 18G2A steels in the range of non-positive stress ratio R (considering elastic-plastic conditions).
- In contrast to J, S^* unambiguously describes the fatigue crack kinetics for 19th-century puddle iron from the Eiffel bridge in the range of positive stress ratio R (considering linear elastic fracture mechanics conditions).
- In each case, the description of KFFD including S^* resulted in higher values of R^2 data fitting coefficient for the power-law description of the FCGR in the Paris regime.
- A good physical interpretation of S^* as opposed to H allows for its easy implementation into the numerical environment.

Funding: This research was funded by Wroclaw University of Science and Technology, grant number 0402/0084/18.

Conflicts of Interest: The funders had no role in the design of the study; in the collection, analyses, or interpretation of data; in the writing of the manuscript; or in the decision to publish the results.

References

1. Paris, P.C.; Erdogan, F. A critical analysis of crack propagation laws. *J. Basic Eng.* **1963**, *85*, 528–533. [CrossRef]
2. Forman, R.G.; Kearney, V.E.; Engle, R.M. Numerical analysis of crack propagation in cyclic loaded structures. *J. Basic Eng.* **1967**, *89*, 459–463. [CrossRef]
3. Walker, K. The effect of stress ratio during crack propagation and fatigue for 2024-T3 and 7075-TO Aluminum. *ASTM Int.* **1970**, *462*, 1–14.
4. Elber, W. Fatigue crack closure under cyclic tension. *Eng. Fract. Mech.* **1970**, *2*, 37–45.
5. Forman, R.G.; Mettu, S.R. Behavior of Surface and Corner Cracks Subjected to Tensile and Bending Loads in Ti–6Al–4V Alloy. In *Fracture Mechanics: 22nd Symposium*; Ernst, H.A., Saxena, A., McDowell, D.L., Eds.; American Society for Testing and Materials: Philadelphia, PA, USA, 1992; Volume 1, pp. 519–546.
6. Koçak, M. FITNET fitness-for-service procedure: An overview. *Weld. World* **2007**, *51*, 94–105. [CrossRef]
7. Huffman, P.J.; Ferreira, J.; Correia, J.A.F.O.; De Jesus, A.M.P.; Lesiuk, G.; Berto, F.; Glinka, G. Fatigue crack propagation prediction of a pressure vessel mild steel based on a strain energy density model. *Frat. Struct. Integrity* **2017**, *11*, 74–84. [CrossRef]
8. Shi, K.K.; Cai, L.X.; Chen, L.; Wu, S.C.; Bao, C. Prediction of fatigue crack growth based on low cycle fatigue properties. *Int. J. Fatigue* **2014**, *61*, 220–225. [CrossRef]
9. Chen, L.; Cai, L.; Yao, D. A new method to predict fatigue crack growth rate of materials based on average cyclic plasticity strain damage accumulation. *Chin. J. Aeronaut.* **2013**, *26*, 130–135. [CrossRef]
10. Boljanović, S.; Maksimović, S.; Djurić, M. Analysis of crack propagation using the strain energy density method. *Sci. Tech. Rev.* **2009**, *59*, 12–17.
11. Khelil, F.; Aour, B.; Belhouari, M.; Benseddiq, N. Modeling of fatigue crack propagation in aluminum alloys using an energy based approach. *Eng. Tech. Appl. Sci. Res.* **2013**, *3*, 488–496.
12. Hadi Hafezi, M.; Nik Abdullah, N.; Correia, J.F.; De Jesus, A.M. An assessment of a strain-life approach for fatigue crack growth. *Int. J. Struct. Integrity* **2012**, *3*, 344–376. [CrossRef]
13. Correia, J.A.; Jesus, A.M.D.; Ribeiro, A.S.; Fernandes, A.A. Strain-based approach for fatigue crack propagation simulation of the 6061-T651 Aluminium alloy. *Int. J. Mater. Struct. Integrity* **2017**, *11*, 1–15. [CrossRef]
14. Zhu, S.P.; Liu, Y.; Liu, Q.; Yu, Z.Y. Strain energy gradient-based LCF life prediction of turbine discs using critical distance concept. *Int. J. Fatigue* **2018**, *113*, 33–42. [CrossRef]
15. Kujawski, D. A fatigue crack driving force parameter with load ratio effects. *Int. J. Fatigue* **2001**, *23*, 239–246. [CrossRef]
16. Kujawski, D. A new ($\Delta K + K_{max}$) 0.5 driving force parameter for crack growth in Aluminum alloys. *Int. J. Fatigue* **2001**, *23*, 733–740. [CrossRef]
17. Stoychev, S.; Kujawski, D. Analysis of crack propagation using ΔK and K_{max}. *Int. J. Fatigue* **2005**, *27*, 1425–1431. [CrossRef]
18. Dinda, S.; Kujawski, D. Correlation and prediction of fatigue crack growth for different R-ratios using Kmax and $\Delta K+$ parameters. *Eng. Fract. Mech.* **2004**, *71*, 1779–1790. [CrossRef]
19. Ranganathan, N.; Chalon, F.; Méo, S. Some aspects of the energy based approach to fatigue crack propagation. *Int. J. Fatigue* **2008**, *30*, 1921–1929. [CrossRef]
20. Mazari, M.; Benguediab, M.; Zemri, M.; Bouchouicha, B. Influence of Structural Parameters on the Resistance on the Crack of Aluminium Alloy. In *Aluminium Alloys-New Trends in Fabrication and Applications*; InTech: London, UK, 2012.
21. Szata, M.; Lesiuk, G. A new method of constructing the kinetic fatigue fracture diagrams - crack propagation equation based on energy approach. *FME Trans.* **2008**, *36*, 74–80.
22. Szata, M.; Lesiuk, G. Algorithms for the estimation of fatigue crack growth using energy method. *Arch. Civ. Mech. Eng.* **2009**, *9*, 118–134. [CrossRef]

23. Szata, M. *Description of the Fatigue Fracture Development in Terms of Energy Approach*; Oficyna Wydawnicza PWr: Wrocław, Poland, 2002. (In Polish)
24. Lesiuk, G.; Szata, M.; Rozumek, D.; Marciniak, Z.; Correia, J.; De Jesus, A. Energy response of S355 and 41Cr4 steel during fatigue crack growth process. *J. Strain Anal. Eng. Des.* **2018**, *53*, 663–675. [CrossRef]
25. Ferreira, J.; Correia, J.A.; Lesiuk, G.; González, S.B.; Gonzalez, M.C.R.; de Jesus, A.M.; Fernández-Canteli, A. Pre-Strain Effects on Mixed-Mode Fatigue Crack Propagation Behaviour of the P355NL1 Pressure Vessels Steel. In *ASME 2018 Pressure Vessels and Piping Conference, Prague, Czech Republic, 15–20 July 2018*; American Society of Mechanical Engineers: New York, NY, USA, 2018; p. V06AT06A027.
26. Lesiuk, G.; Katkowski, M.; Correia, J.; Jesus, A.M.D.; Błażejewski, W. Fatigue crack growth rate in CFRP reinforced constructional old steel. *Int. J. Struct. Integr.* **2018**, *9*, 381–395. [CrossRef]
27. Rozumek, D.; Marciniak, Z.; Lesiuk, G.; Correia, J.A.; de Jesus, A.M. Experimental and numerical investigation of mixed mode I+ II and I+ III fatigue crack growth in S355J0 steel. *Int. J. Fatigue* **2018**, *113*, 160–170. [CrossRef]
28. Lesiuk, G.; Kucharski, P.; Correia, J.A.; de Jesus, A.M.P.; Rebelo, C.; da Silva, L.S. Mixed mode (I + II) fatigue crack growth in puddle iron. *Eng. Fract. Mech.* **2017**, *185*, 175–192. [CrossRef]
29. Dowling, N.E.; Begley, J.A. Fatigue crack growth during gross yielding and the J-integral. In *Mechanics of Crack Growth*; ASTM STP 590; American Society for Testing and Materials: Philadelphia, PA, USA, 1976; pp. 82–103.
30. Dowling, N.E. Geometry effects and the J-integral approach to elasticplastic fatigue crack growth. In *Cracks and Fracture*; ASTM STP 601; American Society for Testing and Materials: Philadelphia, PA, USA, 1976; pp. 19–32.
31. Lamba, H.S. The J-integral approach applied to cyclic loading. *Eng. Fract. Mech.* **1975**, *7*, 693–703. [CrossRef]
32. Wüthrich, C. The extension of the J-integral concept to fatigue cracks. *Int. J. Fract.* **1982**, *20*, 35–37. [CrossRef]
33. Wüthrich, C.; Hoffelner, W. Fatigue crack growth at high strain amplitudes. *Mech. Behav. Mater.* **1984**, 911–917.
34. Tanaka, K. The cyclic J-integral as a criterion for fatigue crack growth. *Int. J. Fract.* **1983**, *22*, 91–104. [CrossRef]
35. Gasiak, G.; Rozumek, D. ΔJ-integral range estimation for fatigue crack growth rate description. *Int. J. Fatigue* **2004**, *26*, 135–140. [CrossRef]
36. Rozumek, D.; Macha, E. Elastic–plastic fatigue crack growth in 18G2A steel under proportional bending with torsion loading. *Fatigue Fract. Eng. Mater. Struct.* **2006**, *29*, 135–144. [CrossRef]
37. Rozumek, D.; Macha, E. *Description of the Fatigue Crack Development in Elastic-Plastic Materials with Proportional Torsion-Bending Load*; Oficyna Wydawnicza Politechniki Opolskiej: Opole, Poland, 2006. (In Polish)
38. Rozumek, D. *Mixed Mode Fatigue Cracks of Constructional Materials, Studies and Monographs*; Oficyna Wydawnicza Politechniki Opolskiej: Opole, Poland, 2009; p. 152. (In Polish)
39. Rozumek, D. Survey of formulas used to describe the fatigue crack growth rate. *Mater. Sci.* **2014**, *49*, 723–733. [CrossRef]
40. Joyce, J.A.; Hackett, E.M.; Roe, C. Effects of cyclic loading on the deformation and elastic-plastic fracture behaviour of a cast stainless steel. In *Fracture Mechanics*; Landes, J.D., McCabe, D.E., Boulet, J.A., Eds.; American Society for Testing and Materials: Philadelphia, PA, USA, 1994; Volume 24, pp. 722–741.
41. Kasprzak, W.; Lysik, B.; Rybaczk, M. *Measurement, Dimensions, Invariant Models and Fractals*; Spolom: Lviv, Ukraine, 2004.
42. Rybaczuk, M. Geometrical methods of dimensional analysis in problems of mechanics. *Scientific Papers of the Institute of Materials Science and Applied Mechanics of the Technical University of Wroclaw*, Wrocław, Poland, 1993. (In Polish)
43. Lesiuk, G.; Szata, M.; Correia, J.A.; De Jesus, A.M.P.; Berto, F. Kinetics of fatigue crack growth and crack closure effect in long term operating steel manufactured at the turn of the 19th and 20th centuries. *Eng. Fract. Mech.* **2017**, *185*, 160–174. [CrossRef]
44. Lesiuk, G. Mixed mode (I + II, I + III) fatigue crack growth rate description in P355NL1 and 18G2A steel using new energy parameter based on j-integral approach. *Eng. Fail. Anal.* **2019**. submitted for publication.
45. Breitbarth, E.; Besel, M. Energy based analysis of crack tip plastic zone of AA2024-T3 under cyclic loading. *Int. J. Fatigue* **2017**, *100*, 263–273. [CrossRef]

46. De Oliveira Correia, J.A.F. An integral probabilistic approach for fatigue lifetime prediction of mechanical and structural components. Ph.D. Thesis, Universidade do Porto, Porto, Portugal, 2015.
47. ASTM E647: Standard Test Method for Measurement of Fatigue Crack Growth Rates. In *Annual Book of ASTM Standards*; ASTM: West Conshohocken, PA, USA, 2011.
48. Becker, T.H.; Mostafavi, M.; Tait, R.B.; Marrow, T.J. An approach to calculate the J-integral by digital image correlation displacement field measurement. *Fatigue Fract. Eng. Mater. Struct.* **2012**, *35*, 971–984. [CrossRef]
49. Vavrik, D.; Jandejsek, I. Experimental evaluation of contour J integral and energy dissipated in the fracture process zone. *Eng. Fract. Mech.* **2014**, *129*, 14–25. [CrossRef]

 © 2019 by the author. Licensee MDPI, Basel, Switzerland. This article is an open access article distributed under the terms and conditions of the Creative Commons Attribution (CC BY) license (http://creativecommons.org/licenses/by/4.0/).

Article

First-Principles Study on the Adsorption and Dissociation of Impurities on Copper Current Collector in Electrolyte for Lithium-Ion Batteries

Jian Chen [1,2], Chao Li [1,2], Jian Zhang [1,2], Cong Li [1,2,3], Jianlin Chen [1,2] and Yanjie Ren [1,2,*]

1. School of Energy and Power Engineering, Changsha University of Science & Technology, Changsha 410114, Hunan, China; chenjian5130@163.com (J.C.); lichao460s@163.com (C.L.); zj4343@163.com (J.Z.); liconghntu@126.com (C.L.); cjlinhunu@csust.edu.cn (J.C.)
2. Key Laboratory of Energy Efficiency and Clean Utilization, Education Department of Hunan Province, Changsha University of Science & Technology, Changsha 410114, Hunan, China
3. Guangxi Key Laboratory of Electrochemical Energy Materials, Guangxi University, Nanning 530004, Guangxi, China
* Correspondence: yjren1008@163.com or yjren@csust.edu.cn; Tel.: +86-731-8525-8409

Received: 11 June 2018; Accepted: 19 July 2018; Published: 21 July 2018

Abstract: The copper current collector is an important component for lithium-ion batteries and its stability in electrolyte impacts their performance. The decomposition of $LiPF_6$ in the electrolyte of lithium-ion batteries produces the reactive PF_6, which reacts with the residual water and generates HF. In this paper, the adsorption and dissociation of H_2O, HF, and PF_5 on the Cu(111) surface were studied using a first-principles method based on the density functional theory. The stable configurations of HF, H_2O, and PF_5 adsorbed on Cu(111) and the geometric parameters of the admolecules were confirmed after structure optimization. The results showed that PF_5 can promote the dissociation reaction of HF. Meanwhile, PF_5 also promoted the physical adsorption of H_2O on the Cu(111) surface. The CuF_2 molecule was identified by determining the bond length and the bond angle of the reaction product. The energy barriers of HF dissociation on clean and O-atom-preadsorbed Cu(111) surfaces revealed that the preadsorbed O atom can promote the dissociation of HF significantly.

Keywords: lithium-ion batteries; copper current collector; first-principles method; adsorption

1. Introduction

Since Armand et al. proposed the concept of rechargeable lithium rocking chair batteries in 1972, lithium-ion batteries have been used widely in portable electronic devices, electric cars, and the aerospace and military fields [1–4]. Although lithium-ion batteries exhibit excellent performance under ambient conditions, during cycling and storage their usable capacity decreases and internal resistance increases rapidly at elevated temperatures [5]. Accordingly, numerous attempts have been performed to increase the thermal stability of lithium-ion batteries. Some studies indicated that $LiPF_6$, widely used as an electrolyte in lithium-ion batteries, is one of the important origins for the capacity fade at elevated temperatures. Ravdel et al. [6] analyzed the thermal decomposition of solid $LiPF_6$ and found that $LiPF_6$ thermally decomposes into PF_5 and LiF. PF_5 is a strong Lewis acid and easily reacts with the solid electrolyte interphase (SEI) in Li-ion batteries. Furthermore, the trace amount of water is always contained in electrolytes (≤50 ppm). Yang et al. [7] investigated the thermal stability of the neat $LiPF_6$ salt in the presence of water (300 ppm) in the carrier gas by thermogravimetric analysis (TGA) and on-line Fourier transform infrared (FTIR). The results showed that pure $LiPF_6$ salt is thermally stable up to 107 °C in a dry, inert atmosphere. However, the initial decomposition temperature reduced to 87 °C due to the presence of water (300 ppm) with the formation of POF_3 and HF. Furthermore, the reaction rates between $LiPF_6$ and water in different solvents for Li-ion batteries are in inverse proportion to the order of their dielectric constants [8]. PF_5 can also react with water to form HF, which can react with organic solvent and the SEI layer. In addition, D. Aurbach [9] proposed that the surface chemistry can control the electrochemical behavior of both lithiated carbon anodes and lithiated transition metal electrodes. During cycling and storage, the spontaneous surface reactions between the SEI and acidic species, such as HF and PF_5, are enhanced. That results in an increase in the electrodes' impedance and causes capacity-fading at elevated temperatures [10–12]. Therefore, some reactive additives or new salts are introduced to the electrolyte to weaken the detrimental effect of decomposition of $LiPF_6$, such as Vinylene carbonate (VC) [5], Li disalicilato-borate salt and silica [12]. Experimentally, Lee et al. [5] found that the SEI layer induced from VC is quite stable at elevated temperatures. However, the thermal decomposition of $LiPF_6$ salt is unavoidable despite the addition of VC. In addition, silica [12] and the graphite coating [13] can absorb HF to protect the material surface from corrosion. Nevertheless, there are still some residual acidic substances to corrode the material surface.

Many researchers have focused on the anode/cathode materials, electrolytes, and separators. However, few studies have been made on current collectors, especially the anode current collector, Cu foil. Myung [14] and Shu [15] et al. found that the presence of HF in the electrolyte was essential in the formation of the metal fluoride layer on the oxide layer of the SEI and water is very important for the formation of passive layers on the surface of the copper current collector. Due to the limitations of these experiments, the microscopic mechanism of the corrosion processes on the surface of Cu foil is not clarified. Hence, combined with the existing experimental phenomena and data, the reaction mechanism of the three vital contaminants, H_2O, HF, and PF_5 on the Cu surface, was investigated by the first-principles method based on the density functional theory.

2. Computational Method and Models

The copper crystal is a face-centered cubic (FCC) structure. Space group is FM-3M. The unit cell contains 4 atoms, as shown in Figure 1a. After structure optimization, the copper lattice parameter is 3.68 Å, which is consistent with the experiment value (3.61 Å) [16]. The Cu(111) surface is the most closely packed plane which is the most stable surface. The close-packed (111) surface energy is the lowest [17]. Therefore, the Cu(111) surface was chosen to study. Considering the boundary effect, the clean Cu(111) (2 × 2) surface model with four layer slabs is constructed. Each layer contains 4 atoms, and the vacuum space is set as 15 Å, as shown in Figure 1b.

In the structural optimization of the constructed models, the bottom layer is fixed to simulate a bulk environment and the others are relaxed. The periodic model calculations are performed using the $Dmol^3$ package based on the density functional theory (DFT). This is done by adopting the generalized gradient approximation [18] (GGA) with the Perdew–Wang 91 (PW91) functional [19] as implemented. All-electron Kohn–Sham wave functions are expanded in a Double Numerical basis [20,21] (DND). Brillouin zone integration is performed using Monkhorst–Pack special k-point grids. In order to get the final structure with minimum total energy, the self-consistent field cycle convergence tolerance is 1×10^{-5} eV and the convergence criteria of optimization is $\leq 2.0 \times 10^{-5}$ Ha (1 Ha = 27.2114 eV), 0.005 Ha/Å, and 0.005 Å for energy, force, and displacement, respectively. HF, H_2O, and PF_5 molecules are placed on a clean surface. To calculate the dissociation energy barriers of HF on the clean and preadsorbed O atom Cu(111) surface, the transition state (TS) [22] of the surface transformation was located on the potential energy hypersurface by performing a linear synchronous transit (LST) calculation, combined with a quadratic synchronous transit (QST) calculation [23], and conjugated gradient refinements. Meanwhile, the smearing energy is set as 0.005 Ha to achieve the fast energy convergence.

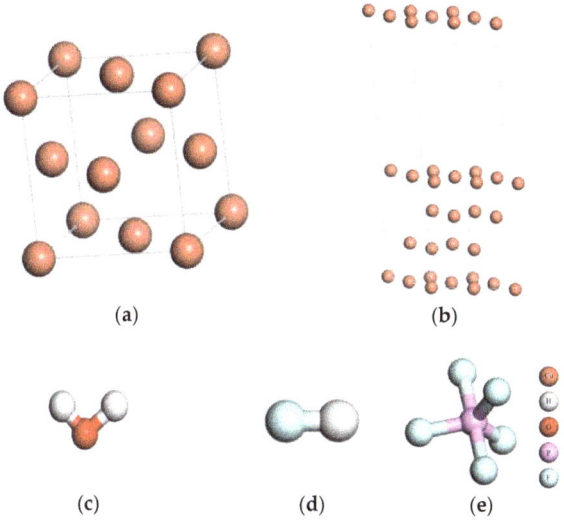

Figure 1. The models of Cu and three kinds of impurities: (**a**) the unit cell, (**b**) the surface periods slab model of the unit cell, (**c**) H_2O, (**d**) HF, and (**e**) PF_5.

3. Results and Discussion

3.1. HF, H_2O, and PF_5 Adsorbed on Clean Cu(111)

Many researchers studied the possible reactions of $LiPF_6$ or water in organic solvents and the following equations have been accepted [24,25]:

$$LiPF_6 \Leftrightarrow LiF + PF_5 \tag{1}$$

$$PF_5 + H_2O \rightarrow POF_3 + 2HF \tag{2}$$

Nonionized $LiPF_6$ dissociates to PF_5, a strong Lewis acid, and LiF in organic solvents. PF_5 reacts with the trace amount of water in the electrolyte and generates POF_3 and HF. PF_5 is highly reactive and sometimes acts as catalyst [6]. Besides that, Zhao et al. [26] experimentally found that even small amounts of impurities, such as H_2O or HF, enhanced the oxidation rate of copper considerably. Thus, the adsorption of H_2O, HF, and PF_5 on the Cu(111) surface are studied in this work.

Figure 2 provides the schematic drawings of HF, H_2O, and PF_5 adsorbed on the Cu(111) surface after structure optimization. As shown in Figure 2a,b, the equilibrium adsorbate–substrate distances of F–Cu and O–Cu are 3.06 Å and 2.46 Å, respectively, which are evidently larger than the sum of the ionic radius of F^-–Cu^{2+} (2.06 Å) and O^{2-}–Cu^{2+} (2.13 Å). That is, the isolated HF or H_2O will not be inclined to adsorb on the surface of Cu(111). According to Figure 2e,f, the equilibrium adsorbate–substrate distances of F–Cu and O–Cu decrease to 2.15 Å and 2.107 Å, respectively, as PF_5 exists simultaneously with HF or H_2O, indicating that PF_5 can promote the adsorption of HF or H_2O.

To further clarify the interaction between HF, H_2O, and PF_5, the geometrical parameters of HF and H_2O with the stable configurations after structure optimization are calculated. Table 1 lists the bond lengths of H–F on the Cu(111) surface in Figure 2a,d,e,g. The bond length of H–F increases from 0.956 Å to 1.009 Å with the coexistence of H_2O and HF, and it increases to 1.345 Å with the existence of HF and PF_5. As PF_5, H_2O, and HF exist simultaneously, the bond length of H–F increases to 2.547 Å. Therefore, both H_2O and PF_5 can promote the dissociation of the H–F bond on the surface of Cu(111). Comparatively, the promotion effect of PF_5 is more pronounced than H_2O.

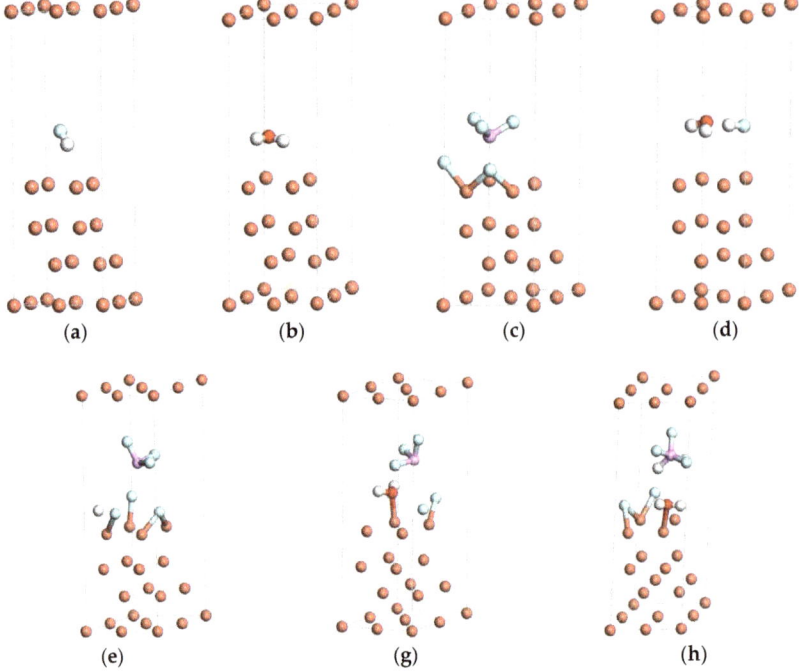

Figure 2. Schematic drawings of HF, H$_2$O, and PF$_5$ adsorbed on the Cu(111) surface after structure optimization: (**a**) HF, (**b**) H$_2$O, (**c**) PF$_5$, (**d**) HF and H$_2$O, (**e**) HF and PF$_5$, (**f**) H$_2$O and PF$_5$, and (**g**) HF, H$_2$O, and PF$_5$.

To clarify the effects of HF, PF$_5$ on the adsorption of H$_2$O, the bond lengths and bond angles of water on the Cu(111) surface are calculated from Figure 2b,d,f,g, which are listed in Table 2. It can be found that the H–O–H internal angle slightly increases from 103.199° to 104.939° with the simultaneous adsorption of H$_2$O and HF. Meanwhile, the internal angle increases to 105.707° with the extra addition of PF$_5$. No evident change was observed for the configuration of H$_2$O molecules, showing that the adsorption of H$_2$O is physical adsorption, which is consistent with the results illustrated by Chen et al. [27].

Table 1. Geometrical parameters of HF molecule on Cu (111) surface in Figure 2.

Condition	(a)	(d)	(e)	(g)
d_{H-F}/Å	0.956	1.009	1.345	2.547

Table 2. Geometrical parameters of H$_2$O molecule on Cu (111) surface in Figure 2.

Condition	(b)	(d)	(f)	(g)
d_{O-H}/Å	0.985	0.985	0.992	0.987
∠(HOH)/(°)	103.199	104.939	105.707	105.255

Figure 3 shows the electron density plots of HF, H$_2$O, and PF$_5$ adsorption on the stable configurations of the Cu(111) surface. As illustrated in Figure 3a,b,d, when HF, H$_2$O, or both of them exist on the Cu(111) surface, there is no obvious overlapping of electron cloud between HF or

H$_2$O and the surface atoms of Cu(111), suggesting that no adsorption occurs for H$_2$O or HF on the surface of copper. However, obvious overlapping of electron cloud could be observed between F atoms from HF and Cu atoms as PF$_5$ and H$_2$O exist simultaneously, as shown in Figure 3e. The similar results could be observed in Figure 3f, as H$_2$O and PF$_5$ coexist. Moreover, the electron contours of F atoms and Cu atoms are smooth and round, indicating that the ionic bond forms between the F atom and Cu atom. It proves that the adsorption of HF is chemisorption. Thus, it can be concluded that PF$_5$ can promote the dissociation of HF and the physical adsorption of H$_2$O on Cu(111) surface. Shu at el. [15] observed P, F, and O elements on the surface of Cu foil which was immersed in the electrolyte of lithium-ion batteries for 30 days and deduced that small amounts of decomposition product (such as PF$_5$) during storage of lithium-ion batteries may have vital effects on the stability of copper. Additionally, some researchers deduced that small amounts of decomposition product (such as PF$_5$) during storage of lithium-ion batteries may have vital effects on the stability of copper [6].

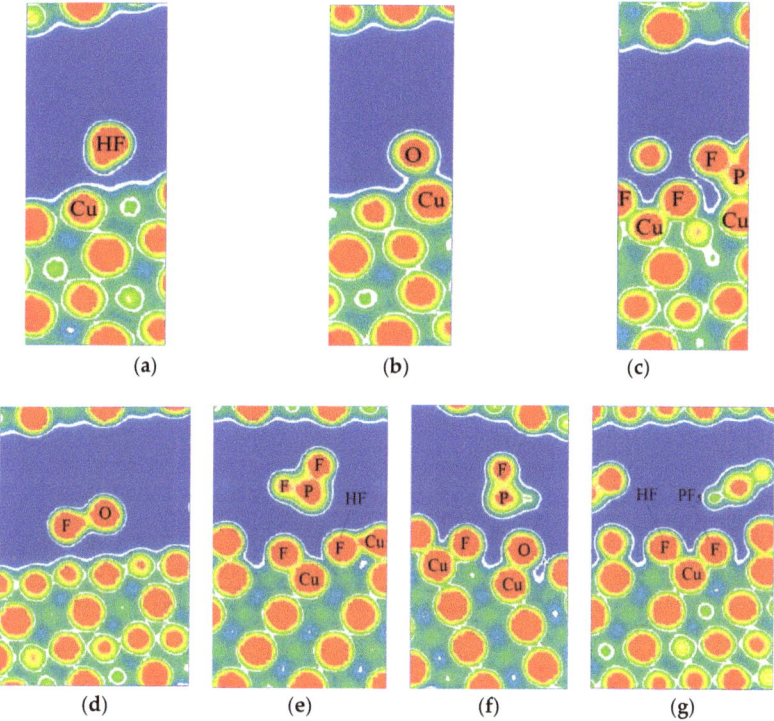

Figure 3. The electron density plots of the stable configurations of HF, H$_2$O, and PF$_5$ adsorbed on Cu(111) surface: (**a**) HF, (**b**) H$_2$O, (**c**) PF$_5$, (**d**) HF and H$_2$O, (**e**) HF and PF$_5$, (**f**) H$_2$O and PF$_5$, and (**g**) HF, H$_2$O, and PF$_5$.

To obtain more information about the interaction between the adsorbates and the surface of Cu(111), the total and partial densities of states (DOS) of related F and Cu atoms are calculated as shown in Figure 4. The Fermi level (E_F) is set as zero and used as a reference. From Figure 4a, the bonding peaks of clean Cu(111) are mainly located at the energy range between E_F and -3 Ha. The bonding electrons between -0.2 and 0 Ha are mainly dominated by the valence electrons of Cu d orbit.

As shown in Figure 4b–e, the total and partial densities of states of Cu(111) with the adsorption of HF, H$_2$O, or both of them are almost identical to the clean Cu(111), which indicates there is no bond formation between HF or H_2O and Cu(111). The results are consistent with those mentioned above. However, a new bonding peak of Cu atoms appears at the low-energy region between −0.2 and −0.15 Ha with addition of PF_5, as shown in Figure 4d,f–h. The peak is stemming from the interaction of Cu 3d orbit and F 2p orbit. The partial densities of states (PDOS) analysis indicates the bond between F and Cu atom is ionic.

According to Figure 4g, the extra addition of PF_5 also results in the expansion of the peak of DOS of F atom between −0.25 and 0 Ha. Meanwhile, a new bonding peak around the energy level of −0.05 Ha appears, which is related to a new hybrid orbit which is generated between Cu atom and F atoms from HF. From Figure 4h, it can be seen that the bonding peaks of F atoms around the energy level of −0.3 Ha almost entirely disappear with the existence of H_2O, HF, and PF_5, and DOS of F atoms is dominated by F 2p orbit. The overlapping of the hybrid orbits is enhanced and the interaction between F and Cu atoms is strengthened. Thus, it can be concluded that the trace amount of H_2O in the electrolyte can promote the spontaneous dissociation of HF and PF_5 on the clean Cu(111) surface.

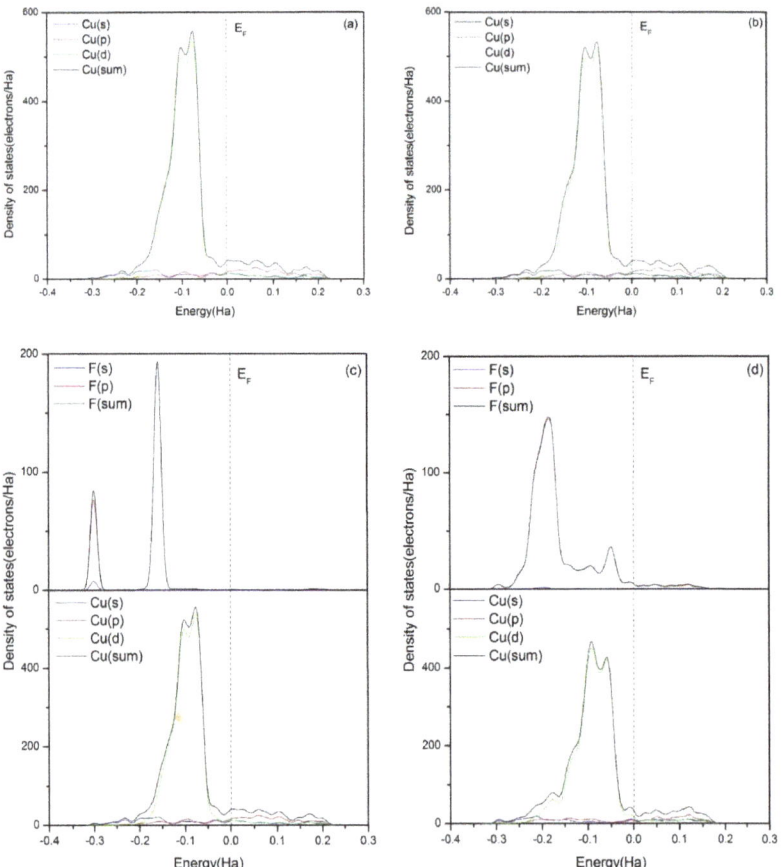

Figure 4. Cont.

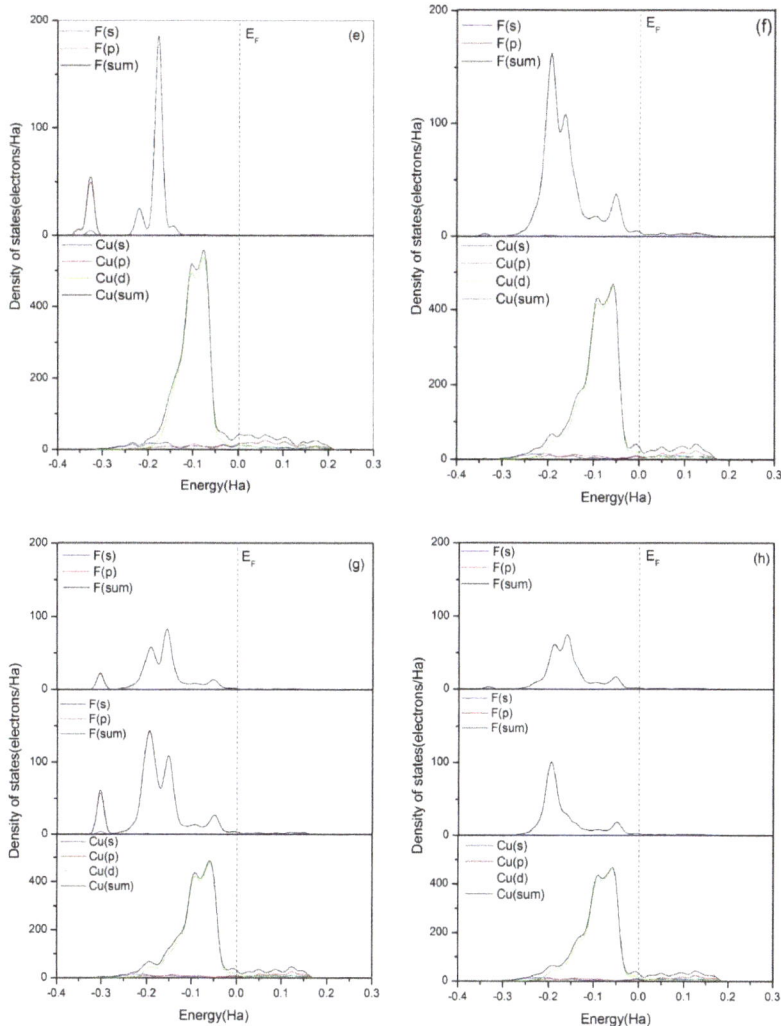

Figure 4. Total and partial density of states of F atom in HF, F atoms in PF$_5$, which form bonds with Cu atoms, and Cu atoms in the topmost layer of clean Cu(111) surface (when the two kinds of F atoms exist together, the F from HF is above). (**a**) Clean Cu(111) surface, (**b**) H$_2$O, (**c**) HF, (**d**) PF$_5$, (**e**) HF and H$_2$O, (**f**) H$_2$O and PF$_5$, (**g**) HF and PF$_5$, and (**h**) HF, H$_2$O, and PF$_5$.

3.2. The Product Formed by F and Cu Atoms

Our previous work shows that CuF$_2$ forms on the surface of copper after immersion in the electrolyte of lithium-ions batteries for 30 days [28]. Myung [14] also reported that metal fluoride was observed in the outer corrosion production layer. According to the lattice parameters of Fischer et al. [29], CuF$_2$ unit cell is constructed to compare with the simulation results. Figure 5a presents the CuF$_2$ unit cell after structure optimization. It can be found that Cu–F bond lengths of CuF$_2$ are 1.929 Å and 1.946 Å, respectively, and the bond internal angle of F–Cu–F is 90.2°. Figure 5b shows the schematic diagram of the bonds formed by the F atoms and Cu atoms as illustrated in Figure 2g. The lengths of Cu–F bonds are 2.103 Å and 2.212 Å and the bond internal angle of F–Cu–F is 89.8°.

The simulation results are consistent with the lattice parameter of the CuF$_2$ molecule. Hence, the results demonstrate theoretically that the reaction product formed by the F and Cu atoms is CuF$_2$.

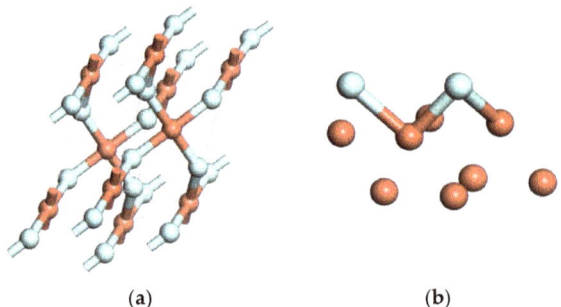

(a)   (b)

Figure 5. Schematic drawing: (**a**) the CuF$_2$ unit cell and (**b**) the bonds formed by the F atoms and Cu atoms in Figure 2g.

3.3. HF Dissociation on Clean and Preadsorbed O/Cu(111) Surfaces

In fact, Shu et al. [15] found that HF etches the copper oxides layer on the copper current collector more severely. To verify this, the dissociation energy barriers of HF on the clean and preadsorbed O/Cu(111) surfaces were calculated. During HF dissociation processes, the most stable adsorption sites of the related atoms should be provided. There are four adsorption sites of H, F, and O atoms adsorbed on the surface, and the adsorption energies are calculated as follows:

$$E_{ad} = E_{a/Cu(111)} - E_{Cu(111)} - E_a \tag{3}$$

where $E_{Cu(111)}$ is the total energy of the clean Cu(111) surface, E_a is the total energy of the adsorbed atom, and $E_{a/Cu(111)}$ is the total energy of the Cu(111) surface with the adsorbed atom. The greater the absolute value of the adsorption energy is, the stronger the interaction between the adsorbed atom and the surface. Table 3 lists the adsorption energies of H, F, and O atoms on clean Cu(111). After structure optimization, it is found that the H, F, and O atoms at the top site are relaxed to the fcc site, and the H, F, and O atoms at the bridge site are relaxed to the hcp site. According to the adsorption energies and the positions of H, F, and O atoms in Figure 6, it is concluded that the fcc site is the preferable adsorption site of H, F, and O atoms on the Cu(111) surface.

According to the calculated results of HF adsorption configurations on the Cu(111) surface, no spontaneous dissociation of HF can be observed. Hence, the activation energy is required for the dissociation reaction. The stable adsorption configurations of HF on two surfaces are chosen as the reactants of dissociation and the stable adsorption configurations of separate H and F at two fcc sites are predicted as the products of HF dissociation. The activation barrier of the dissociation reaction is calculated by locating the transition state with the linear synchronous transit (LST) method. The energy and structural evolution of the systems for HF dissociation on clean and preadsorbed O atom on the Cu(111) surface are shown in Figure 7. It can be seen that the sites of separate H and F adsorbed on the Cu(111) surface are in agreement with that of predicted dissociation product. The energy barrier for the dissociation of HF on the clean Cu(111) surface is 114.27 kJ/mol and it reduces significantly to 19.58 kJ/mol for HF on O atom preadsorbed Cu(111) surface. Hence, the preadsorbed O atom acts as a catalyst and dramatically slashes the required activation energy of the HF dissociation.

The adsorption energies and geometrical parameters of the HF molecule on clean and O preadsorbed Cu(111) surfaces are calculated to explain the reduction of the dissociation energy barrier. From Table 4, it can be seen that the adsorption energy of HF on O/Cu(111) surfaces is 0.52 eV, which is evidently higher than that on clean Cu(111) surfaces (i.e., 0.27 eV), showing that the adsorption of HF on O atom preadsorbed Cu (111) surfaces is more stable than that on clean Cu(111) surfaces. Moreover, the bond length of HF on Cu(111) surfaces with the preadsorbed O is 0.95 Å, which is longer than that on clean Cu(111) surfaces. Thus, it can be concluded that the preadsorbed O atom can promote the break of the H–F bond. Since O atom is more electronegative than F atom, it is apt to break the H–F bond. Thus, it also demonstrates that HF is prone to etching the copper oxides on the surface of copper, as reported in the reference [14,15].

Table 3. Adsorption energies of H, F, and O atoms on Cu(111) surface (eV).

Site	Top	Bridge	Hcp	Fcc
H	−2.619	−2.617	−2.620	−2.622
O	−4.970	−4.846	−4.845	−4.971
F	−4.024	−4.001	−4.001	−4.025

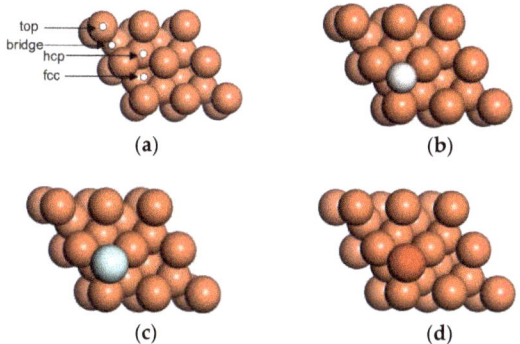

Figure 6. Stable adsorption configuration of adatom on Cu(111) surface: (a) surface adsorption sites of Cu(111), (b) H, (c) F, and (d) O.

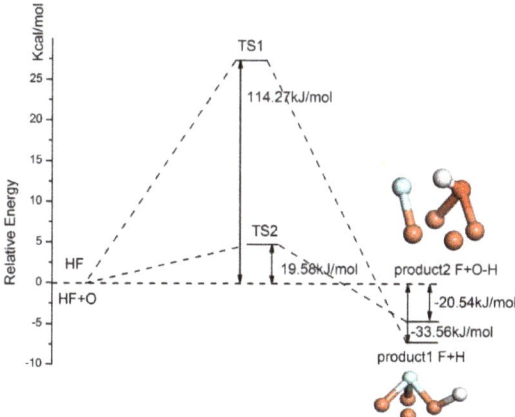

Figure 7. The dissociation pathway for HF on different Cu(111) surfaces.

Table 4. Geometrical parameters and adsorption energies of HF molecule on clean and O atom preadsorbed Cu(111) surfaces.

Surface Type	E_{ad} (eV)	d_{H-F} (Å)
Clean Cu(111)	0.27	0.95
O preadsorbed Cu(111)	0.52	0.98

4. Conclusions

In lithium-ions batteries, PF_5 stemming from the thermal decomposition of electrolytes can react with the residual water in the electrolyte and produce HF. To investigate the effects of PF_5, H_2O, and HF on the stability of the copper current collector of lithium-ions batteries, the adsorption of HF, H_2O, and PF_5 on Cu(111) surfaces was systematically studied based on the density functional theory.

(1) Both H_2O and PF_5 can promote the dissociation of HF. PF_5 also has a promotion effect on the physical adsorption of H_2O on Cu(111) surfaces. Meanwhile, the spans of the DOS of F atom from HF between −0.25 and 0 Ha enlarge obviously and a new bonding peak around the energy level of −0.05 Ha appears with the addition of PF_5, indicating a new hybrid orbit is generated between the F atom from HF and the Cu atom. The bond between the F atom and Cu atom is ionic, and the reaction product is a CuF_2 molecule.

(2) The most stable adsorption sites of the related atoms (H, O, and F) are all the fcc sites. The dissociation energy barrier on O adsorbate-adsorbed Cu(111) surfaces is much less than that on the clean Cu(111) surface. The preadsorbed O atom plays a catalytic role to dramatically slash the required activation energy of the dissociation of HF.

Author Contributions: Writing-original draft, J.C.; Software, C.L.; Data curation, J.Z. and C.L.; Investigation, J.C.; Writing-review & editing, Y.R.

Funding: This research was funded by the Natural Science Foundation of China, grant number 51471036 and 51771034.

Conflicts of Interest: The authors declare no conflict of interest. The funders had no role in the design of the study; in the collection, analyses, or interpretation of data; in the writing of the manuscript; and in the decision to publish the results.

References

1. Hong, L.I. Fundamental scientific aspects of lithium ion batteries (XV)—Summary and outlook. *Energy Storage Sci. Technol.* **2015**, *4*, 306–318.
2. Waag, W.; Fleischer, C.; Sauer, D.U. Critical review of the methods for monitoring of lithium-ion batteries in electric and hybrid vehicles. *J. Power Sources* **2014**, *258*, 321–339. [CrossRef]
3. Ratnakumar, B.V.; Smart, M.C.; Kindler, A.; Frank, H.; Ewell, R.; Surampudi, S. Lithium batteries for aerospace applications: 2003 Mars Exploration Rover. *J. Power Sources* **2003**, *119*, 906–910. [CrossRef]
4. Lee, H.; Yanilmaz, M.; Toprakci, O.; Fu, K.; Zhang, X. A review of recent developments in membrane separators for rechargeable lithium-ion batteries. *Energy Environ. Sci.* **2014**, *7*, 3857–3886. [CrossRef]
5. Lee, H.H.; Wang, Y.Y.; Wan, C.C.; Yang, M.H.; Wu, H.C.; Shieh, D.T. The function of vinylene carbonate as a thermal additive to electrolyte in lithium batteries. *J. Appl. Electrochem.* **2005**, *35*, 615–623. [CrossRef]
6. Ravdel, B.; Abraham, K.M.; Gitzendanner, R.; DiCarlo, J.; Lucht, B.; Campion, C. Thermal stability of lithium-ion battery electrolytes. *J. Power Sources* **2003**, *119*, 805–810. [CrossRef]
7. Yang, H.; Zhuang, G.V.; Ross, P.N. Thermal stability of $LiPF_6$ salt and Li-ion battery electrolytes containing $LiPF_6$. *J. Power Sources* **2006**, *161*, 573–579. [CrossRef]
8. Kawamura, T.; Okada, S.; Yamaki, J. Decomposition reaction of LiPF 6-based electrolytes for lithium ion cells. *J. Power Sources* **2006**, *156*, 547–554. [CrossRef]
9. Aurbach, D. Electrode–solution interactions in Li-ion batteries: A short summary and new insights. *J. Power Sources* **2003**, *119*, 497–503. [CrossRef]

10. Koltypin, M.; Aurbach, D.; Nazar, L.; Ellis, B. On the stability of LiFePO4 olivine cathodes under various conditions (electrolyte solutions, temperatures). *Electrochem. Solid-State Lett.* **2007**, *10*, A40–A44. [CrossRef]
11. Myung, S.T.; Sasaki, Y.; Saito, T.; Sun, Y.K.; Yashiro, H. Passivation behavior of Type 304 stainless steel in a non-aqueous alkyl carbonate solution containing $LiPF_6$, salt. *Electrochim. Acta* **2009**, *54*, 5804–5812. [CrossRef]
12. Philippe, B.; Dedryvère, R.; Gorgoi, M.; Rensmo, H.; Gonbeau, D.; Edström, K. Role of the LiPF6 Salt for the Long-Term Stability of Silicon Electrodes in Li-Ion Batteries—A Photoelectron Spectroscopy Study. *Chem. Mater.* **2013**, *25*, 394–404. [CrossRef]
13. Zhao, M.; Xu, M.; Dewald, H.D.; Staniewicz, R.J. Open-Circuit Voltage Study of Graphite-Coated Copper Foil Electrodes in Lithium-Ion Battery Electrolytes. *J. Electrochem. Soc.* **2003**, *150*, A117–A120. [CrossRef]
14. Myung, S.T.; Sasaki, Y.; Sakurada, S.; Sun, Y.K.; Yashiro, H. Electrochemical behavior of current collectors for lithium batteries in non-aqueous alkyl carbonate solution and surface analysis by ToF-SIMS. *Electrochim. Acta* **2009**, *55*, 288–297. [CrossRef]
15. Shu, J.; Shui, M.; Huang, F.; Xu, D.; Ren, Y.; Hou, L.; Cui, J.; Xu, J. Comparative study on surface behaviors of copper current collector in electrolyte for lithium-ion batteries. *Electrochim. Acta* **2011**, *56*, 3006–3014. [CrossRef]
16. Straumanis, M.E.; Yu, L.S. Lattice parameters, densities, expansion coefficients and perfection of structure of Cu and of Cu-In phase. *Acta Crystallogr. Sect. A Cryst. Phys. Diffr. Theor. Gen. Crystallogr.* **1969**, *25*, 676–682. [CrossRef]
17. Tafreshi, S.S.; Roldan, A.; Dzade, N.Y.; de Leeuw, N.H. Adsorption of hydrazine on the perfect and defective copper (111) surface: A dispersion-corrected DFT study. *Surf. Sci.* **2014**, *622*, 1–8. [CrossRef]
18. Blöchl, P.E. Projector augmented-wave method. *Phys. Rev. B* **1994**, *50*, 17953. [CrossRef]
19. Perdew, J.P.; Burke, K.; Ernzerhof, M. Generalized gradient approximation made simple. *Phys. Rev. Lett.* **1996**, *77*, 3865. [CrossRef] [PubMed]
20. Delley, B. Analytic energy derivatives in the numerical local-density-functional Approach. *J. Chem. Phys.* **1991**, *94*, 7245–7250. [CrossRef]
21. Luo, Y.; Yin, S.; Lai, W.; Wang, Y. Effects of global orbital cutoff value and numerical basis set size on accuracies of theoretical atomization energies. *Theor. Chem. Acc.* **2014**, *133*, 1580. [CrossRef]
22. Henkelman, G.; Uberuaga, B.P.; Jónsson, H. A climbing image nudged elastic band method for finding saddle points and minimum energy paths. *J. Chem. Phys.* **2000**, *113*, 9901–9904. [CrossRef]
23. Halgren, T.A.; Lipscomb, W.N. The synchronous-transit method for determining reaction pathways and locating molecular transition states. *Chem. Phys. Lett.* **1977**, *49*, 225–232. [CrossRef]
24. Heider, U.; Oesten, R.; Jungnitz, M. Challenge in manufacturing electrolyte solutions for lithium and lithium ion batteries quality control and minimizing contamination level. *J. Power Sources* **1999**, *81*, 119–122. [CrossRef]
25. Aurbach, D.; Zaban, A.; Ein-Eli, Y.; Weissman, I.; Chusid, O.; Markovsky, B.; Levi, M.; Levi, E.; Schechter, A.; Granot, E. Recent studies on the correlation between surface chemistry, morphology, three-dimensional structures and performance of Li and Li-C intercalation anodes in several important electrolyte systems. *J. Power Sources* **1997**, *68*, 91–98. [CrossRef]
26. Zhao, M.; Kariuki, S.; Dewald, H.D.; Lemke, F.R.; Staniewicz, R.J.; Plichta, E.J.; Marsh, R.A. Electrochemical Stability of Copper in Lithium-Ion Battery Electrolytes. *J. Electrochem. Soc.* **2000**, *147*, 2874–2879. [CrossRef]
27. Chen, J.; Li, C.; Ren, Y.J.; Chen, J.L. First-principles study of adsorption and dissociation of H2O on Cu(111) surface. *J. Changsha Universi. Sci. Technol. Natural Sci.* **2017**, *14*, 92–97.
28. Dai, S.W.; Chen, J.; Ren, Y.J.; Liu, Z.M.; Chen, J.L.; Li, C.; Zhang, X.Y.; Zhang, X.; Zeng, T.F. Electrochemical Corrosion Behavior of Copper Current Collector in Electrolyte for Lithium-ion Batteries. *Int. J. Electrochem. Sci.* **2017**, *12*, 10589–10598. [CrossRef]
29. Fischer, P.; Hälg, W.; Schwarzenbach, D.; Gamsjäger, H. Magnetic and crystal structure of copper(II) fluoride. *J. Phys. Chem. Solids* **1974**, *35*, 1683–1689. [CrossRef]

© 2018 by the authors. Licensee MDPI, Basel, Switzerland. This article is an open access article distributed under the terms and conditions of the Creative Commons Attribution (CC BY) license (http://creativecommons.org/licenses/by/4.0/).

Article

Surface Nanocrystallization and Amorphization of Dual-Phase TC11 Titanium Alloys under Laser Induced Ultrahigh Strain-Rate Plastic Deformation

Sihai Luo [1], Liucheng Zhou [1], Xuede Wang [1], Xin Cao [1], Xiangfan Nie [1,3,*] and Weifeng He [1,2,*]

[1] Science and Technology on Plasma Dynamics Laboratory, Air Force Engineering University, Xi'an 710038, China; luo_hai@126.com (S.L.); happyzlch@163.com (L.Z.); wangxuede@163.com (X.W.); studentcaoxin@163.com (X.C.)
[2] Institute of Aeronautics Engine, School of Mechanical Engineering, Xi'an Jiaotong University, Xi'an 710049, China
[3] School of Mechanical and Power Engineering, East China University of Science and Technology, Shanghai 200237, China
* Correspondence: skingkgd@163.com (X.N.); hehe_coco@163.com (W.H.); Tel.: +86-029-8478-7537 (X.N.); +86-029-8478-7537 (W.H.)

Received: 7 March 2018; Accepted: 3 April 2018; Published: 6 April 2018

Abstract: As an innovative surface technology for ultrahigh strain-rate plastic deformation, laser shock peening (LSP) was applied to the dual-phase TC11 titanium alloy to fabricate an amorphous and nanocrystalline surface layer at room temperature. X-ray diffraction, transmission electron microscopy, and high-resolution transmission electron microscopy (HRTEM) were used to investigate the microstructural evolution, and the deformation mechanism was discussed. The results showed that a surface nanostructured surface layer was synthesized after LSP treatment with adequate laser parameters. Simultaneously, the behavior of dislocations was also studied for different laser parameters. The rapid slipping, accumulation, annihilation, and rearrangement of dislocations under the laser-induced shock waves contributed greatly to the surface nanocrystallization. In addition, a 10 nm-thick amorphous structure layer was found through HRTEM in the top surface and the formation mechanism was attributed to the local temperature rising to the melting point, followed by its subsequent fast cooling.

Keywords: laser shock peening; dual-phase TC11 titanium alloy; ultrahigh strain-rate plastic deformation; nanocrystallization; amorphization

---

1. Introduction

Titanium alloys are the most widely utilized alloy in the aero-engine industry. Due to their high fatigue strength, low density, and high corrosion resistance, they are employed in components such as disks, fans, and compressor blades. Nonetheless, the fatigue failure of titanium alloy blades, especially those subjected to foreign object damage, has been a major concern [1,2]. In order to improve the fatigue resistance, various surface modification techniques, such as mechanical attrition treatment (SMAT), high-pressure torsion, and shot peening, have been proposed to improve the mechanical properties and fatigue strength of titanium alloys [2–4]. One mechanism of fatigue strength improvement is to induce surface nanocrystallization, the main reason is that nanocrystalline materials have many exceptional physical, mechanical, and fatigue resistance properties, relative to their coarse-grained counterparts [5–10]. However, the drawback of these proposed processing techniques is that they are only suitable for the simple components, which restricted their industrial application to aero-engine blades.

Compared with other surface treatment technologies, laser shock peening (LSP) have the remarkable advantages of not requiring contact with the substrate, of affecting a deep layer, and of having an excellent controllability, which makes LSP very suitable for complex structures, for example, aero-engine blades [11–13]. Similar to conventional shot peening, LSP is capable of refining grains and producing a surface nanostructured layer [14–26]. Recently, LSP has been used on the aero-engine components to improve the high cycle fatigue performance and to induce surface nanocrystallization of stainless steels [18,19], titanium alloys [20–22], and nickel-based superalloys [23,24] has been reported. Lu et al. [14–16] discussed, in detail, the mechanism of grain refinement induced by LSP on LY2 aluminum alloy, AISI 304 stainless steel, and commercially pure titanium. Luo et al. [25] further analyzed the mechanism of surface nanocrystallization induced by LSP on the metallic alloys with different stacking fault energies. Moreover, Lainé et al. [26] analyzed, in detail, the effects of metallic shot peening (MSP) and LSP on the microstructure of Ti-6Al-4V. The results showed that the grain refinement of MSP was the evolution of tangled dislocation structures and shear bands, whereas that of LSP was the evolution of directional planar dislocations and networks of dislocation cells and sub-grains. In addition, the formation of an amorphous structure was noticed on the NiTi shape memory alloy and silicon after laser-induced shock wave compression [27,28]. To summarize, LSP is beneficial to the microstructural change of the surface layer and thus, improves the fatigue strength. Thus, it is of great interest to investigate the microstructural evolution mechanism of the TC11 titanium alloy under LSP treatment.

In this work, the microstructures of dual-phase TC11 titanium alloys treated by LSP were characterized by transmission electron microscopy (TEM). The microstructural evolution rule and surface nanocrystallization mechanism were investigated. In addition, a special phenomenon of surface amorphization on the top layer was observed by high-resolution transmission electron microscopy (HRTEM), and the formation mechanism was also discussed.

2. Experimental Procedures

2.1. Materials

TC11 titanium alloy is typically employed for fan blades in the Chinese aero-engine industry. The chemical composition (in wt %) is given in Table 1. The sample was an α + β type heat-resistant titanium alloy, composed of a primary hexagonal close-packed (hcp) α phase and a lamellar transformed bcc β phase. The equiaxial α-grains with an average size of 10 μm and the acicular β-grains were observed by the optical microscopy, as shown in Figure 1. The alloy was subjected to a double annealing heat treatment for 2 h at 950–980 °C followed by air cooling, and heating for 6 h at 530 °C, followed by 6 h of air cooling. The basic mechanical properties of these titanium alloys are shown in Table 2.

Figure 1. An optical micrograph of the TC11 titanium alloy (etched for ~12 s in a solution of 5% HNO$_3$ and 95% of absolute alcohol).

Table 1. The chemical composition of the TC11 titanium alloy.

Composition	Al	Mo	Cr	Zr	Si	Fe	Sn	Ti
Percentage (wt %)	5.8–7.0	2.8–3.8	–	0.8–2.0	0.15–0.40	0.2–0.7	–	Bal

Table 2. The static tensile properties of the TC11 titanium alloy at room temperature.

Materials	Yield Strength $\sigma_{0.2}$ (MPa)	Ultimate Tensile Strength σ_b (MPa)	Elongation Rate δ (%)
TC11 titanium alloy	930	1030	9

2.2. Principle and Experimental Procedure of LSP

In the LSP process, a laser pulse with a short pulse width (ns) and a high power density (GW/cm^2) was placed on the workpiece surface. The workpiece to be laser peened was covered by two different layers, namely an opaque ablating layer (Al foil/black paint) and a transparent confining layer (water/glass), as shown in Figure 2. The laser pulse passed through the transparent confining medium and struck the ablating layer. It was then absorbed by the ablating layer, which immediately vaporized and formed plasmas of a high temperature and pressure. The expansion of the plasma detonation wave led to the formation of a shock wave that propagated into the target with an intensity of several GPa. When the shockwave pressure was larger than the material dynamic yield strength, a plastic deformation was produced and resulted in the generation of compressive residual stresses and microstructural changes in the material surface layer.

Before LSP treatment, the sample surface was polished with SiC paper with a grit number ranging between 400 and 2000. An ultrasound ethanol bath was then used to clean the surface of the sample. During the LSP experiment, the confining layer and ablating layers consisted of floating water with a thickness of about 1–2 mm and a 0.1 mm-thick Al foil, respectively. The titanium alloys were machined into square samples (40 mm × 40 mm × 4 mm) which were mounted on a five-coordinate robot arm. The robot arm was controlled to move in the x-y direction in order to achieve the designed laser paths, shown in Figure 2. The laser pulse with a wavelength of 1064 nm and a pulse of around 20 ns was generated by a Q-switched self-designed Nd:YAG laser (SGR-EXTRA/25J). The laser spot diameter, overlapping-rate, and repletion-rate were 3 mm, 50%, and 1 Hz, respectively. In order to investigate the effect of laser parameters on the microstructural characteristics, different laser power densities (2.83, 4.24, and 5.66 GW/cm^2) at three impacts were adopted.

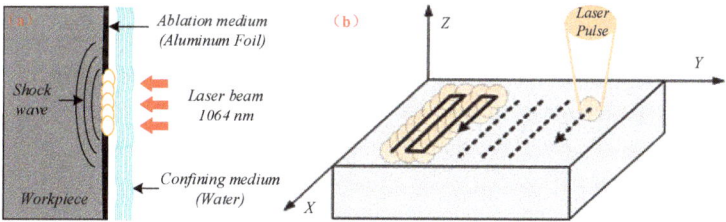

Figure 2. Schematic illustrations of the laser shock peening process. (**a**) The plasma shock wave generated by nanosecond pulse laser; (**b**) The LSP processed area of samples for microstructural observation and laser shock paths (the LSP processed area had the dimensions of 25 × 20 mm).

2.3. Microstructural Observations

X-ray diffraction (XRD) analyses of the TC11 titanium alloy with and without LSP treatment were conducted using an MFS-7000 X-ray diffractometer with Cu-Kα radiation (Shimadzu, Kyoto, Japan). The take-off angle was 6° and the generator settings were 40 kV and 35 mA. The diffraction data were collected for values of 2θ ranging from 30° to 80° at a step of 0.02° and a time step of 5 s.

TEM HRTEM observations on the LSP-treated samples were performed using a JEM-2100F (Japan Electron Optics Laboratory, Beijing, China) with the following experimental parameters: FEG (field emission gun): 200 kV; point resolution: 0.23 nm; and line resolution: 0.14 nm. The TEM foils for the surface layers of the samples were prepared by mechanically grinding the samples on the side that had not been subjected to LSP in order to obtain thin plates with a thickness of about 50 μm. The thin plates were then electro-polished using a twin-jet technique in a liquid solution consisting of 300 mL of CH_3OH, 175 mL of C_4H_9OH, and 30 mL of $HClO_4$ (30% solubility). The grain size measurements were made directly from the dark-field TEM images. In addition, the focused ion beam (FIB) lift-out method was used to prepare the cross-sectional TEM samples from the top surface of the LSP-treated sample in the FEI Helios NanoLab™ 600i system, and the HRTEM observation was used to analyze the microstructural characteristics at different depths.

3. Results and Discussion

3.1. Microstructure Characterization on the Surface

Figure 3 shows the XRD patterns of the LSP-treated specimens with different laser power densities at three impacts. After LSP treatment, the Bragg diffraction peak of the TC11 titanium alloy became broader and decreased in intensity, which indicated that the grain refinement, lattice deformation, and micro-strain increases had been induced on the surface layer of the alloy. It is worth noting that, as the power density increased, the Bragg diffraction peaks broadened more significantly and eventually flattened out. On the other hand, the XRD spectral peak and position remained virtually unchanged, indicating that no phase change had occurred. The ultrahigh strain rate plastic deformation was induced during the LSP process, which resulted in the generation of a non-uniform residual elastic micro-strain in the materials and a microstructural change. This was the reason for the tendency of the broadening of the diffraction peak to deviate towards low angles, as seen in the inset of Figure 3.

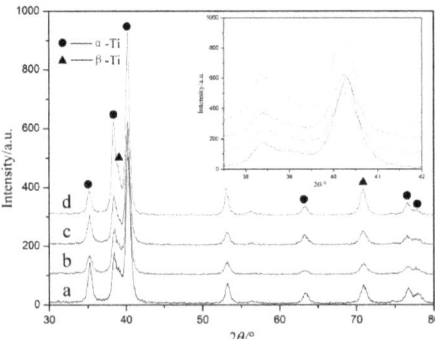

Figure 3. The X-ray diffraction pattern of the surface microstructure with different laser power density at three impacts. (**a**) original; (**b**) 2.83 GW/cm²; (**c**) 4.24 GW/cm²; (**d**) 5.66 GW/cm². The inset shows the XRD pattern for 2θ from 37.5° to 42°.

The XRD patterns showed that the laser power density had a direct influence on the microstructural change following LSP treatment. The XRD measurement results were relative to the surface layer with a depth of about 1 μm (the penetrated depth of the X-ray). Thus, in order to further confirm the grain refinement induced by the LSP treatment and to analyze the microstructural evolution of the TC11 titanium alloy under ultrahigh strain-rate deformation, TEM observations were carried out.

Figure 4 shows the TEM images obtained from the top surface layer after LSP treatment. The original TC11 titanium alloy was composed of α and β phases and the phase boundary, as seen in

Figure 4a. The greatest dimension of these phases reached several micrometers. After LSP with a low laser power density (2.83 GW/cm^2), high-density dislocation configurations (dislocation, dislocation tangle, and dislocation cell) were generated near the grain boundaries, as shown in Figure 4b. When the laser power density increased to 4.24 GW/cm^2, many nanocrystalline artifacts were generated, as shown in Figure 4c,d. The corresponding selected area electron diffraction (SAED) pattern was dominated by circles, which indicated the random orientations of the nanocrystalline artifacts and their high angle grain boundaries. When the power density was increased to 5.66 GW/cm^2, the surface nanocrystallization was completed, as shown in Figure 4e,f, and the grain size was refiner and more uniform compared to that induced by lower laser power densities. The corresponding SAED pattern presented continuous, homogeneous and broadened concentric rings, confirming the random crystallographic orientation of the grains. In a previous work, we investigated the effect of multiple LSP impacts on the TC11 titanium alloy at a power density of 4.24 GW/cm^2 [29] and found that the high-density dislocations and dislocation walls were formed after one impact, and the numerous nanocrystalline artifacts were generated after three LSP impacts. When the number of LSP impacts increased to five, the average grain size decreased to about 40 nm and the grain orientation of the nanocrystalline artifacts became more random and uniform. Therefore, increasing either the laser power density or the LSP impacts was effective in inducing surface nanocrystallization. In other words, grain refinement increases with the amount of laser energy injected into the materials.

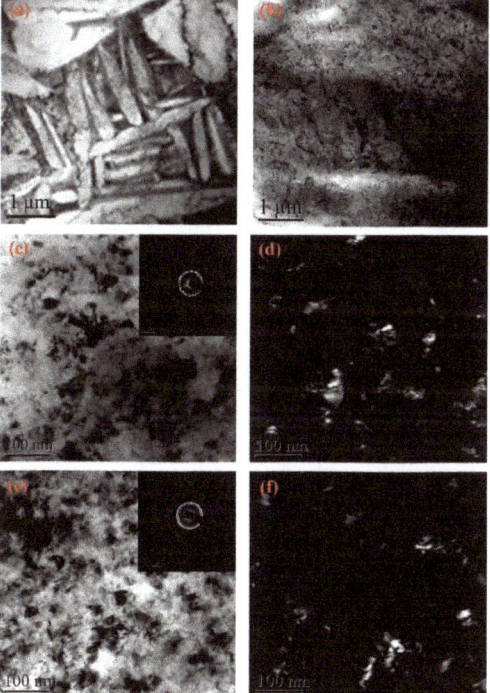

Figure 4. The transmission electron microscopy TEM images and corresponding diffraction patterns of the surface layer of the TC11 titanium alloy samples after LSP with different laser power densities at three impacts. (**a**) without LSP; (**b**) 2.83 GW/cm^2; (**c**) 4.24 GW/cm^2, the inset is the corresponding SAED pattern; (**d**) the corresponding dark-field image of (c); (**e**) 5.66 GW/cm^2, the inset is the corresponding SAED pattern; (**f**) the corresponding dark-field of (e).

To investigate the microstructural characteristics at different depths of TC11 titanium alloys after LSP treatment. The chosen sample was the one that had been subjected to LSP with a power density of 4.24 GW/cm^2 at three impacts. The cross-sectional microstructure of TC11 titanium alloy is shown in Figure 5. The top region in Figure 5a was a carbon (C) deposition layer which was used to protect the sample from the ion beam during the TEM sample preparation by FIB. The integrity of the C deposition layer showed that the sample had not been damaged during the preparation process. The SAED pattern in the selected region, A, showed that the nanostructure was produced after LSP treatment, consistent with the results reported in Figure 5. Beneath the nanocrystalline layer, at a depth of about 350 nm, the slight elongation of the corresponding SAED pattern points indicated to the presence of high-density dislocations and sub-structures in the selected region, B. The microstructural characteristics at different depths were consistent with the attenuation rule of the laser-induced shock wave pressure in the materials. The SAED pattern of the top surface (region C) presented halo ring characteristics, which may have been caused by the presence of either very fine grains or amorphous phases. To clarify the microstructural morphology, HRTEM observations were carried out, as shown in Figure 5b, which confirmed the amorphous structure of the material for 10 nm in thickness. For depths greater than 10 nm, the microstructure was composed of both nanocrystalline and amorphous phases. This was also confirmed by the corresponding SAED pattern of region C, in which both halo rings and diffraction spots were presented.

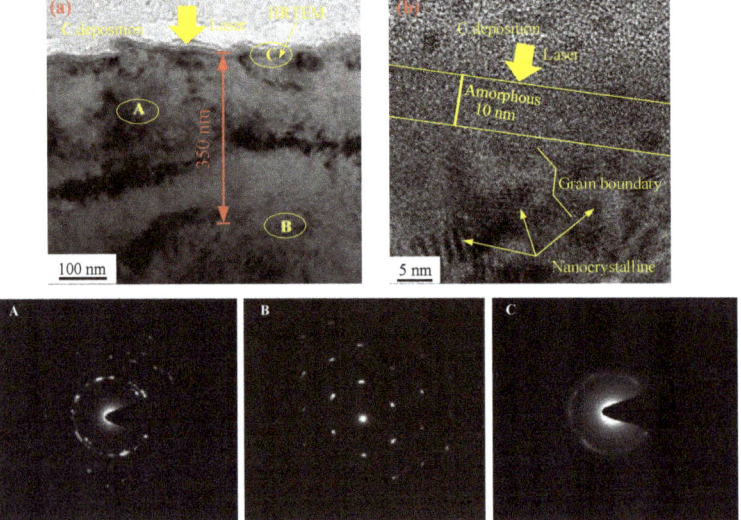

Figure 5. TEM photographs and selected diffraction patterns in cross-section and HRTEM photographs on the top surface layer of TC11 titanium alloy treated by LSP with three impacts. (**a**) The cross-sectional TEM image, and the plastic deformation layer is divided into three layers, typically regions, A, B and C). (**b**) The HRTEM observation of region C in (**a**).

3.2. Surface Nanocrystallization

The results described in Section 3.1 show that different microstructures were generated after LSP treatment with different laser parameters, including high-density dislocations, dislocation tangles, dislocation cells, and nano-grains. Hence, the surface nanocrystallization process was similar to the one caused by conventional severe plastic deformation methods, such as shot peening and surface mechanical attrition treatment [3,6]. On the other hand, there are many differences in the surface nanocrystallization mechanism between LSP and conventional shot peening.

In conventional shot peening, the plastic deformation with a strain-rate of about 10^3 s^{-1} occurs in the contact region when the material surface is struck by hard particles, and the surface nanocrystallization is a stepwise evolution process. As for LSP, when the shock wave pressure induced by LSP was larger than the material dynamic yield strength, plastic deformation occurred and the strain-rate of the plastic deformation reached 10^6 s^{-1}. During the plastic deformation process, the atoms were forced to move because of the laser-induced shock wave. According to the homogeneous nucleation theory [30,31], the atoms usually moved in arrays at the shock wavefront, which led to the formation of dislocations. Unlike the dislocation formation process during conventional shot peening, these dislocations were generated at the shock wavefront in the direction of the shock wave propagation. If dislocations were generated, the subsequent shock wave caused the dislocations to ship; if the dislocations met each other or other crystal defects, the dislocation slipping ceased. The dislocations, therefore, gather in correspondence of special locations, where there is a great resistance to their movement, giving rise to the formation of two special statuses: dislocation tangles and dislocation cells. The process of microstructural change was completed extremely rapidly, which was attributed to the fact that the action of the shock wave lasted for only a few tens of nanoseconds. As shown in Figure 4 and in the previous work [29], the surface nanocrystallization was realized when the laser power density or the LSP impacts increased. An increase in the laser power density or LSP impact corresponds to an increase in the laser energy injection into the materials, or an increase in the dynamic plastic deformation time of the laser-induced shock wave, as described in Figure 6. The injected laser energy was transformed into material plasticity energy by the action of the shock wave and was stored in the crystal defects such as dislocations and grain boundaries. The degree of grain refinement increases with the laser energy injected in the material. Therefore, when the dynamic plastic deformation time of the shock wave increases, the dislocations were further driven to slip. Lastly, the dislocation cells transformed into nano-grains, while the dislocation walls transformed into nano-grain boundaries by the annihilation and rearrangement of dislocations. On the other hand, due to the local temperature rising during the LSP process (detailed discussion in Section 3.3), the dynamic recrystallization took place. Therefore, the surface nanocrystallization mechanism consisted of three main steps: (i) the formation of high-density dislocations; (ii) the pileup of dislocation to dislocation cells; (iii) the formation of sub-grain boundaries and of surface nanocrystalline artifacts through continuous energy injection and dynamic recrystallization.

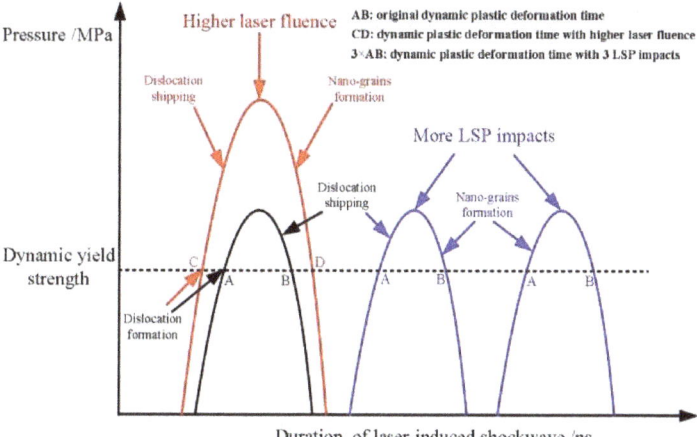

Figure 6. The principle of the action of the shockwave for increasing the laser power density or LSP impact. t_{AB} and t_{CD} are the plastic deformation times at a fixed power density and at a higher power density, respectively; 3-AB is the dynamic plastic deformation time with the three impacts.

3.3. Mechanism of Laser-Induced Amorphization

Amorphization is considered as an extreme case of grain refinement. There are two methods for synthesizing amorphous structures in metallic materials [32,33]: (i) "freezing" the dynamic disorder of a liquid using an extremely high cooling rate, that is, rapid quenching; (ii) destabilizing and "melting" a solid to the amorphous state by introducing chemical and/or structural disorder, which is known as solid-state amorphization. For example, Ye et al. [27] found that an amorphous phase could be generated by LSP and discussed the formation mechanism based on the plasticity theory. Similar results and formation mechanisms were reported by Wang et al. [34], who found that the threshold peak pressure for amorphization of a NiTi alloy was about 3.3 GPa (the pressure is 4.4 GPa in this case). Meyers [35,36] found that shear bands were generated in AISI 304 stainless steel and germanium under shock loading conditions, and the formation of an amorphous phase was observed in these shear bands. The liquid-solid structure induced by the local temperature rising and subsequent fast cooling were the main reasons for the formation of an amorphous phase; however, the LSP-induced surface amorphization of titanium alloys has not been reported in the literature so far. Therefore, it is necessary to investigate the mechanism of surface amorphization by LSP on titanium alloys.

According to the results reported in References [34–37], the crystal-to-amorphous transition can be attributed to the local temperature rising to the melting point of the material due to plastic deformation and its subsequent fast cooling. Worswick and Yang [38] assumed that 5% of the plastic deformation work was stored in the grain defects and 95% was transformed into heat. In addition, the duration of the plastic deformation induced by LSP was only a few tens of nanoseconds. Thus, this thermodynamic process at the shock wavefront was regarded as an adiabatic process. The adiabatic temperature rising during the course of the dynamic loading could be calculated as [39] follows:

$$\Delta T = \frac{\beta}{\rho C_p} \int_0^{\varepsilon_f} \sigma d\varepsilon \tag{1}$$

where β is the Taylor–Quinney coefficient which characterizes the portion of plastic deformation work converted into heat (assumed to be 0.9 in this case), ρ is the density (4.48 g/cm^3 for the present TC11 titanium alloy), C_p is the specific heat capacity (0.48 J/(g·K) for the present TC11 titanium alloy), σ is the flow stress, and ε_f is the true strain in the final state. The dynamic mechanical behavior in the plastic deformation process is usually described by the Johnson–Cook constitutive model and the flow stress σ induced by LSP is expressed by Equation (2) [40]. Thus, the temperature rising could be expressed by Equation (3) after substituting the Johnson–Cook constitutive equation.

$$\sigma = (\sigma_0 + B\varepsilon^n)\left(1 + C \ln \frac{\dot{\varepsilon}}{\dot{\varepsilon}_0}\right) \tag{2}$$

$$\Delta T = \frac{0.9(1 + C \ln \dot{\varepsilon}/\dot{\varepsilon}_0)}{\rho C_p}\left(\sigma_0 \varepsilon_f + \frac{B\varepsilon_f^{n+1}}{n+1}\right) \tag{3}$$

where σ_0 and $\dot{\varepsilon}_0$ are the initial field strength and reference strain-rate. B, C, and n are the constants of the Johnson–Cook constitutive equation. For the titanium alloys, the relative constants are as follows: σ_0 is 1030 MPa (shown in Table 2), $\dot{\varepsilon}_0$ is 10^{-2}/s, and $\dot{\varepsilon}$ is 10^7/s; B, C, and n are 1092 MPa, 0.014, and 0.93, respectively [41]; and ε_f is about 2 for the local strain. The local temperature rising ΔT was found to be about 1990 K, which was higher than the melting temperature T_m (about 1800 K) of the TC11 titanium alloy.

During the LSP process, once the temperature in the material reached T_m, melting occurred without consuming any enthalpy of fusion. The temperature rising in the melting region transited into the neighboring regions to the melting point, until the heat flow from the melting zone was no longer sufficient to raise the temperature of its neighbor to T_m. The cooling rate in the melting region was at least of the order of 10^7 K/s [42], which resulted in the liquid-to-amorphous phase transformation. This is why a 10 nm-thick of amorphous phase was observed by HRTEM in the present TC11 titanium alloy.

3.4. Microhardness Distribution

To better characterize the base material behavior to LSP, changes in the properties were evaluated via Vickers microhardness measurements. In this study, the microhardness of samples was tested by the MVS-1000JMT2 microhardness tester (BAHENS, Shanghai, China), using an indentation load of 500 g with a dwell time of 15 s at the section. For each depth of the specimen, the hardness value was regarded as an average of 5 measurement results and a confidence interval of 95%.

The cross-sectional microhardness curves of TC11 titanium alloy after LSP treatment is shown in Figure 7. It is observed that the average value of surface microhardness of the original specimen is approximately 351 HV0.5. After LSP treatment, the microhardness was effectively increased and the maximum value was located at the surface at 424 HV0.5. The affected depth is about 500 μm and the gradient change of the microhardness is consistent with the attenuation of the laser-induced shock wave pressure. The microhardness improvement is attributed to the surface work hardening and microstructural changes induced by LSP. The relationship between microhardness and microstructure can be expressed by the Hall–Petch equation [43]:

$$H_v = H_0 + kd^{-1/2} + \alpha Gb\rho^{1/2} \qquad (4)$$

where H_v is the microhardness of the material, H_0 is the original hardness of the material, k and α are material constants, d is the grain size, G is the shear modulus of the material, b is the Burgers vector, and ρ is the dislocation density. After multiple LSP treatments, the grain sizes in the surface layer are refined and high-density dislocation is found in the substrate layer. According to the Hall–Petch model, the microhardness increases after LSP treatment. Due to the thickness being only 10 nm, the effect of the surface amorphousness on microhardness cannot be assessed.

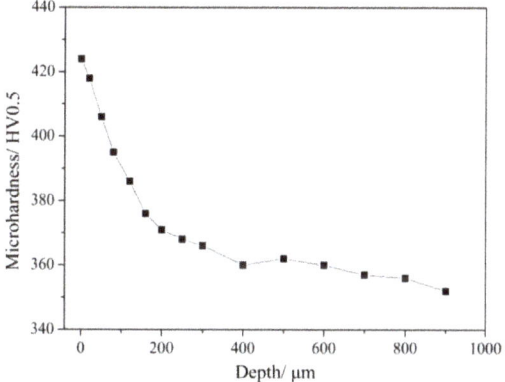

Figure 7. The cross-sectional microhardness of the TC11 titanium alloy with LSP treatment.

4. Conclusions

In this paper, the surface and cross-sectional microstructures of the TC11 titanium alloy were characterized via transmission electron microscopy and high-resolution transmission electron microscopy. According to the different microstructural features, the mechanisms of surface nanocrystallization and amorphization after LSP treatment were discussed. The main conclusions obtained in this work may be summarized as follows:

(1) Surface nanocrystallization was induced by LSP on the TC11 titanium alloy. In the LSP process, the dislocations were generated at the shock wavefront. They then developed into dislocation cells and finally formed the nano-grains by dislocation, slipping under the continuous shock wave, and dynamic recrystallization.

(2) More specially, an amorphous layer of about 10 nm thickness was generated on the top surface above the nanostructured layer. The local temperature rising during the LSP process resulted from the dynamic compression and ultrahigh strain-rate plastic deformation under the laser-induced high-pressure shock wave. The combined effect of the temperature rising to the melting point and the fast cooling caused the surface amorphization of the TC11 titanium alloy.

Acknowledgments: This work is sponsored by the National Basic Research Program of China (No. 2015CB057400), the National Natural Science Foundation of China (No. 51505496 and 51405507) and the National Postdoctoral Program for Innovative Talents (No. 201700077), China Postdoctoral Science Foundation (No. 191785).

Author Contributions: Weifeng He and Xiangfan Nie conceived and designed the experiments; Weifeng He, Xin Cao, and Liucheng Zhou performed the experiments; Sihai Luo, Xin Cao, and Xiangfan Nie analyzed the data and wrote the paper; Xuede Wang checked the paper. All authors read and approved the manuscript.

Conflicts of Interest: The authors declare no conflict of interest.

References

1. Zabeen, S.; Preuss, M.; Withers, P.J. Evolution of a laser shock peened residual stress field locally with foreign object damage and subsequent fatigue crack growth. *Acta Mater.* **2015**, *83*, 216–226. [CrossRef]
2. Liu, W.C.; Wu, G.H.; Zhai, C.Q.; Ding, W.J.; Korsunsky, A.M. Grain refinement and fatigue strengthening mechanisms in as-extruded Mg–6Zn–0.5Zr and Mg–10Gd–3Y–0.5Zr magnesium alloys by shot peening. *Int. J. Plasticity* **2013**, *49*, 16–35. [CrossRef]
3. Wang, N.; Peña, L.V.W.; Wang, L.; Mellor, B.G.; Huang, Y. Experimental and Simulation Studies of Strength and Fracture Behaviors of Wind Turbine Bearing Steel Processed by High Pressure Torsion. *Energies* **2016**, *9*, 1033. [CrossRef]
4. Zhang, K.; Wang, Z.B.; Lu, K. Enhanced fatigue property by suppressing surface cracking in a gradient nanostructured bearing steel. *Mater. Res. Lett.* **2017**, *5*, 258–266. [CrossRef]
5. Lu, K.; Lu, J. Surface nanocrystallization (SNC) of metallic materials-presentation of the concept behind a new approach. *J. Mater. Sci. Technol.* **1999**, *15*, 193–197.
6. Laleh, M.; Kargar, F. Effect of surface nanocrystallization on the microstructural and corrosion characteristics of AZ91D magnesium alloy. *J. Alloys Compd.* **2011**, *509*, 9150–9156. [CrossRef]
7. Lu, K.; Lu, J. Nanostructured surface layer on metallic materials induced by surface mechanical attrition treatment. *Mater. Sci. Eng. A* **2004**, *375–377*, 38–45. [CrossRef]
8. Tao, N.R.; Wang, Z.B.; Tong, W.P.; Sui, M.L.; Lu, J.; Lu, K. An investigation of surface nanocrystallozation mechanism in Fe induced by surface mechanical attrition treatment. *Acta Mater.* **2002**, *50*, 4603–4616. [CrossRef]
9. Wang, K.; Tao, N.R.; Liu, G.; Lu, J.; Lu, K. Plastic strain-induced grain refinement at the nanometer scale in copper. *Acta Mater.* **2006**, *54*, 975–982. [CrossRef]
10. Zhang, H.W.; Hei, Z.K.; Liu, G.; Lu, J.; Lu, K. Formation of nanostructured surface layer on AISI 304 stainless steel by means of surface mechanical attrition treatment. *Acta Mater.* **2003**, *51*, 1871–1881. [CrossRef]
11. Zhang, L.; Lu, J.Z.; Zhang, Y.K.; Ma, H.L.; Luo, K.Y.; Dai, F.Z. Effects of laser shock processing on morphologies and mechanical properties of ANSI 304 stainless steel weldments subjected to cavitation erosion. *Materials* **2017**, *10*, 292. [CrossRef] [PubMed]
12. Montross, C.S.; Wei, T.; Ye, L.; Clark, G.; Mai, Y.W. Laser shock processing and its effects on microstructure and properties of metal alloys: A review. *Int. J. Fatigue* **2002**, *24*, 1021–1036. [CrossRef]
13. Ye, C.; Suslov, S.; Kim, B.J.; Stach, E.A.; Cheng, G.J. Fatigue performance improvement in AISI 4140 steel by dynamic strain aging and dynamic precipitation during warm laser shock peening. *Acta Mater.* **2011**, *59*, 1014–1025. [CrossRef]
14. Lu, J.Z.; Luo, K.Y.; Zhang, Y.K.; Sun, G.F.; Gu, Y.Y.; Zhou, J.Z.; Ren, X.D.; Zhang, X.C.; Zhang, L.F.; Chen, K.M.; et al. Grain refinement mechanism of multiple laser shock processing impacts on ANSI 304 stainless steel. *Acta Mater.* **2010**, *16*, 5354–5362. [CrossRef]
15. Lu, J.Z.; Luo, K.Y.; Zhang, Y.K.; Cui, C.Y.; Sun, G.F.; Zhou, J.Z.; Zhang, L.; You, J.; Chen, K.M.; Zhong, J.W. Grain refinement of LY2 aluminum alloy induced by ultra-high plastic strain during multiple laser shock processing impacts. *Acta Mater.* **2010**, *58*, 3984–3994. [CrossRef]

16. Lu, J.Z.; Wu, L.J.; Sun, G.F.; Luo, K.Y.; Zhang, Y.K.; Cai, J.; Cui, C.Y.; Luo, X.M. Microstructural response and grain refinement mechanism of commercially pure titanium subjected to multiple laser shock peening impacts. *Acta Mater.* **2017**, *127*, 252–266. [CrossRef]
17. Luo, S.H.; He, W.F.; Zhou, L.C.; Nie, X.F.; Li, Y.H. Aluminizing mechanism on a nickel-based alloy with surface nanostructure produced by laser shock peening and its effect on fatigue strength. *Surf. Coat. Technol.* **2018**, *342*, 29–36.
18. Zhou, L.; He, W.F.; Wang, X.D.; Zhou, L.C.; Li, Q.P. Effect of laser shock processing on high cycle fatigue properties of 1Cr11Ni2W2MoV stainless steel. *Rare Metal Mater. Eng.* **2011**, *40*, 174–177.
19. Zhou, L.C.; He, W.F.; Luo, S.H.; Long, C.B.; Wang, C.; Nie, X.F.; He, G.Y.; Shen, X.J.; Li, Y.H. Laser shock peening induced surface nanocrystallization and martensite transformation in austenitic stainless steel. *J. Alloys Compd.* **2016**, *655*, 66–70. [CrossRef]
20. Zhou, L.C.; Li, Y.H.; He, W.F.; He, G.Y.; Nie, X.F.; Chen, D.L.; Lai, Z.L.; An, Z.B. Deforming TC6 titianium alloys at ultrahigh strain rates during multiple laser shock peening. *Mater. Sci. Eng. A* **2013**, *578*, 181–186. [CrossRef]
21. Nie, X.F.; He, W.F.; Zhou, L.C.; Li, Q.P.; Wang, X.D. Experiment investigation of laser shock peening on TC6 titanium alloy to improve high cycle fatigue performance. *Mater. Sci. Eng. A* **2014**, *594*, 161–167. [CrossRef]
22. Ren, X.D.; Zhou, W.F.; Liu, F.F.; Ren, Y.P.; Yuan, S.Q.; Ren, N.F.; Xu, S.D.; Yang, T. Microstructure evolution and grain refinement of Ti-6Al-4V alloy by laser shock processing. *Appl. Surf. Sci.* **2016**, *363*, 44–49. [CrossRef]
23. Luo, S.H.; Nie, X.F.; Zhou, L.C.; You, X.; He, W.F.; Li, Y.H. Thermal stability of surface nanostructure produced by laser shock peening in a Ni-based superalloy. *Surf. Coat. Technol.* **2017**, *311*, 337–343.
24. Li, Y.H.; Zhou, L.C.; He, W.F.; He, G.Y.; Wang, X.D.; Nie, X.F.; Wang, B.; Luo, S.H.; Li, Y.Q. The strengthening mechanism of a nickel-based alloy after laser shock processing at high temperatures. *Sci. Technol. Adv. Mater.* **2013**, *14*, 055010. [CrossRef] [PubMed]
25. Luo, S.H.; Li, Y.H.; Zhou, L.C.; Nie, X.F.; He, G.Y.; Li, Y.Q.; He, W.F. Surface nanocrystallization of metallic alloys with different stacking fault energy induced by laser shock processing. *Mater. Des.* **2016**, *104*, 320–326.
26. Lainé, S.J.; Knowles, K.M.; Doorbar, P.J.; Cutts, R.D.; Rugg, D. Microstructural characterisation of metallic shot peened and laser shock peened Ti–6Al–4V. *Acta Mater.* **2017**, *123*, 350–361. [CrossRef]
27. Ye, C.; Suslov, S.; Fei, X.L.; Cheng, G.J. Bimodal nanocrystallization of NiTi shape memory alloy by laser shock peening and post-deformation annealing. *Acta Mater.* **2011**, *59*, 7219–7227. [CrossRef]
28. Zhao, S.; Hahn, E.N.; Kad, B.; Remington, B.A.; Wehrenberg, C.E.; Bringa, E.M.; Meyers, M.A. Amorphization and nanocrystallization of silicon under shock compression. *Acta Mater.* **2016**, *103*, 519–533. [CrossRef]
29. Nie, X.F.; He, W.F.; Zang, S.L.; Wang, X.D.; Zhao, J. Effect study and application to improve high cycle fatigue resistance of TC11 titanium alloy by laser shock peening with multiple impacts. *Surf. Coat. Technol.* **2014**, *253*, 68–75. [CrossRef]
30. Meyers, M.A.; Subhash, G.; Kad, B.K.; Prasad, L. Evolution of microstructure and shear-band formation in α-hcp titanium. *Mech. Mater.* **1994**, *17*, 175–193. [CrossRef]
31. Meyers, M.A.; Gregori, F.; Kad, B.K.; Schneider, M.S.; Kalantar, D.H.; Remington, B.A.; Ravichandran, G.; Boehly, T.; Wark, J.S. Laser-induced shock compression of monocrystalline copper: Characterization and analysis. *Acta Mater.* **2003**, *51*, 1211–1228. [CrossRef]
32. Han, S.; Zhao, L.; Jiang, Q.; Lian, J.S. Deformation-induced localized solid-state amorphization in nanocrystalline nickel. *Sci. Rep.* **2012**, *2*, 493. [CrossRef] [PubMed]
33. Enayati, M.H.; Mohamed, F.A. Application of mechanical alloying/milling for synthesis of nanocrystalline and amorphous materials. *Int. Mater. Rev.* **2014**, *59*, 394–416. [CrossRef]
34. Wang, X.; Xi, W.G.; Wu, X.Q.; Wei, Y.P.; Huang, C.G. Microstructure and mechanical properties of an austenite NiTi shape memory alloy treated with laser induced shock. *Mater. Sci. Eng. A* **2013**, *578*, 1–5. [CrossRef]
35. Meyers, M.A.; Xu, Y.B.; Xue, Q.; Pérez-Prado, M.T.; McNelley, T.R. Microstructural evolution in adiabatic shear localization in stainless steel. *Acta Mater.* **2003**, *51*, 1307–1325. [CrossRef]
36. Zhao, S.T.; Kad, B.; Wehrenberg, C.E.; Remington, B.A.; Hahn, E.N.; More, K.L.; Meyers, M.A. Generating gradient germanium nanostructures by shock-induced amorphization and crystallization. *Proc. Natl. Acad. Sci. USA* **2017**, *114*, 9791–9796. [CrossRef] [PubMed]
37. Peterlechner, M.; Waitz, T.; Karnthaler, H.P. Nanoscale amorphization of severely deformed NiTi shape memory alloys. *Scripta Mater.* **2009**, *60*, 1137–1140. [CrossRef]

38. Worswick, M.J.; Qiang, N.; Niessen, P.; Pick, R.J. *Shock Wave and High Strain Rate Phenomena in Materials*; Marcel Dekker Inc.: New York, NY, USA, 1992.
39. Liu, J.X.; Li, S.K.; Zhou, X.Q.; Zhang, Z.H.; Zheng, H.Y.; Wang, Y.C. Adiabatic shear banding in a tungsten heavy alloy processed by hot-hydrostatic extrusion and hot torsion. *Scripta Mater.* **2008**, *59*, 1271–1274. [CrossRef]
40. Ding, K.; Ye, L. *Laser Shock Peening Performance and Process Simulation*; Woodhead: New York, NY, USA, 2006.
41. Lesuer, D.R. *Final Report: DOT/FAA/AR-00/25*; US Department of Transportation, Federal Aviation Administration: Washington, WA, USA, 2000.
42. Li, N.; Wang, Y.D.; Peng, R.L.; Sun, X.; Liaw, P.K.; Wu, G.L.; Wang, L.; Cai, H.N. Localized amorphism after high-strain-rate deformation in TWIP steel. *Acta Mater.* **2011**, *59*, 6369–6377. [CrossRef]
43. Zhang, Z.; Chen, D.L. Consideration of Orowan strengthening effect in particulate-reinforced metal matrix nanocomposites: A model for predicting their yield strength. *Scripta Mater.* **2006**, *54*, 1321–1326. [CrossRef]

 © 2018 by the authors. Licensee MDPI, Basel, Switzerland. This article is an open access article distributed under the terms and conditions of the Creative Commons Attribution (CC BY) license (http://creativecommons.org/licenses/by/4.0/).

Article

Indentation Behavior and Mechanical Properties of Tungsten/Chromium co-Doped Bismuth Titanate Ceramics Sintered at Different Temperatures

Shaoxiong Xie [1], Jiageng Xu [2], Yu Chen [3,4,*], Zhi Tan [3], Rui Nie [3], Qingyuan Wang [1,4,*] and Jianguo Zhu [3]

1. College of Architecture and Environment, Sichuan University, Chengdu 610065, China; xsxdyx@126.com
2. School of Architecture and Civil Engineering, Chengdu University, Chengdu 610106, China; xxujiageng@163.com
3. College of Materials Science and Engineering, Sichuan University, Chengdu 610065, China; tanzhi0838@163.com (Z.T.); nierui129@163.com (R.N.); nic0400@scu.edu.cn (J.Z.)
4. School of Mechanical Engineering, Chengdu University, Chengdu 610106, China
* Correspondence: chenyuer20023@163.com (Y.C.); wangqy@scu.edu.cn (Q.W.)

Received: 26 February 2018; Accepted: 22 March 2018; Published: 27 March 2018

Abstract: A sort of tungsten/chromium(W/Cr) co-doped bismuth titanate (BIT) ceramics ($Bi_4Ti_{2.95}W_{0.05}O_{12.05}$ + 0.2 wt % Cr_2O_3, abbreviate to BTWC) are ordinarily sintered between 1050 and 1150 °C, and the indentation behavior and mechanical properties of ceramics sintered at different temperatures have been investigated by both nanoindentation and microindentation technology. Firstly, more or less $Bi_2Ti_2O_7$ grains as the second phase were found in BTWC ceramics, and the grain size of ceramics increased with increase of sintering temperatures. A nanoindentation test for BTWC ceramics reveals that the testing hardness of ceramics decreased with increase of sintering temperatures, which could be explained by the Hall–Petch equation, and the true hardness could be calculated according to the pressure-state-response (PSR) model considering the indentation size effect, where the value of hardness depends on the magnitude of load. While, under the application of microsized Vickers, the sample sintered at a lower temperature (1050 °C) gained four linearly propagating cracks, however, they were observed to shorten in the sample sintered at a higher temperature (1125 °C). Moreover, both the crack deflection and the crack branching existed in the latter. The hardness and the fracture toughness of BTWC ceramics presented a contrary variational tendency with increase of sintering temperatures. A high sintering tends to get a lower hardness and a higher fracture toughness, which could be attributed to the easier plastic deformation and the stronger crack inhibition of coarse grains, respectively, as well as the toughening effect coming from the second phase.

Keywords: $Bi_4Ti_3O_{12}$ ceramics; sintering temperature; crack propagation; mechanical properties; indentation behavior

---

1. Introduction

Bismuth layered structure ferroelectrics (BLSF), as a kind of deuterogenic perovskite compounds with a high Curie temperature, have become a competitive candidate for the sensitive materials of those piezoelectric/ferroelectric devices with high operating temperatures [1] in view of their interesting electromechanical-coupling behaviors and fatigue-free properties [2]. In the family of BLSFs, $Bi_4Ti_3O_{12}$ (BIT) as a typical member has attracted great interest during the 1970s because of its high Curie temperatures of ~675 °C and large spontaneous polarization of ~50 µC/cm² along the a-axis. BIT has been reported to have considerable potential for application in some high-temperature

(300 °C~400 °C) piezoelectric transducers after its high electrical conductivity were significantly decreased by Nb-doping [3].

In most piezoelectric sensors and actuators, ferroelectric ceramics are prone to fatigue due to cyclic electrical or mechanical loadings. The fatigue manifests its effect as a reduction in domain switching ability or mechanical strength and subsequent premature failure of devices. However, such mechanical properties of ferroelectric ceramics, which significantly influences the reliability of the devices. These often neglected because we used to pay more attention to their electrical properties, which are more relevant to the output of devices. Especially in some severe working environments involved with high temperature, strong coupling and high frequency, etc. [4,5], some complicated mechanical behavior of ferroelectric ceramics including fatigue crack propagation, creep deformation, and brittle–ductile transition have to be considered in the structural design of devices for the guarantee of reliability. To characterize the mechanical behavior of ferroelectric ceramics is of paramount importance in understanding their in-service failure mechanisms based on the knowledge that the sharp indenter has considerable potential as a microprobe for quantitatively characterizing mechanical properties. The indentation technology has been widely used in the last three decades for measuring the mechanical properties of small-scale materials such as electrical ceramics [6,7]. On the other hand, it is well known that the sintering temperatures plays an important role in the microstructural development of ceramics, further influencing its macroscopic mechanical properties [8–11].

Recently, W/Cr co-doped $Bi_4Ti_3O_{12}$ ceramics were identified to have a low electric conductivity and a high piezoelectric constant [12,13]. However, there is hardly any report referring to the influence of sintering temperatures on the mechanical properties of this material. In this paper, a sort of W/Cr co-doped $Bi_4Ti_3O_{12}$ ceramics with the optimal chemical composition were synthesized by a traditional ceramic process. We investigated the microstructural evolution of ceramics with the sintering temperatures, revealing the correlation between the deformation mechanism and microstructures of ceramics by two mechanical testing including nanoindentation and Vickers indentation.

2. Experiment

2.1. Preparation of Ceramics

A sort of W/Cr co-doped $Bi_4Ti_3O_{12}$ ceramics with a chemical formula of $Bi_4Ti_{2.95}W_{0.05}O_{12.05}$ + 0.2 wt % Cr_2O_3 (abbreviated as BTWC), were fabricated by two steps using the conventional solid-state reaction technique. First of all, reagent-grade oxide powders: Bi_2O_3 (99.999%), TiO_2 (98%) and WO_3 (99%) (Sinopharm Chemmical Reagent Co., Ltd., Shanghai, China) were weighed in the stoichiometric amounts ($Bi_4Ti_{2.95}W_{0.05}O_{12.05}$) of the ceramics. These raw materials were mixed by planetary ball mill using ethanol as solvent and zirconia as grinding balls for 24 h. This homogeneous mixture was calcined at 850 °C for 4 h to synthesize the compound of $Bi_4Ti_{2.95}W_{0.05}O_{12.05}$ after drying. Secondly, 0.2 wt % of Cr_2O_3 (99%) was added into the calcined powders and then mixed with them in the same method. The dried powders were granulated with polyvinyl alcohol (PVA, 8%). And then, the powders were compacted into discs with a diameter of 10 mm and a thickness of 1 mm under an isostatic stress of 150 MPa. After PVA was burned out at 450 °C, these discs were sintered at a temperature range of 1050–1150 °C for 4 h in a sealed alumina crucible to get BTWC ceramics.

2.2. Characterization of Ceramics

2.2.1. Microstructural Characteristics

The actual density of ceramics was measured by the Archimedes method. The phase structures of ceramics were determined by an X-ray diffractometer (XRD, DX2700, Dandong, China) using Cu-Kα radiation (λ = 1.5418 Å) at room temperature. The microstructural morphology of the ceramics was observed by scanning electron microscopy (SEM, JSM-610LV, JEOL, Tokyo, Japan) focusing on their natural surfaces. The average grain size was obtained by the linear intercept method from the SEM images.

2.2.2. Nanoindentation Test

A nanoindentation testing system (Hysitron Triboscope, Hysitron, Eden Prairie, MN, USA) conducted by a Berkovich indenter was employed to investigate the elastic and plastic properties of the ceramics. Firstly, the surfaces of ceramics were finely polished using diamond pastes. In all tests, both the loading rate and the unloading rate were maintained at 0.5 mN/s for each peak load (50 mN, 100 mN, 150 mN and 200 mN). Forces and displacements were synchronously recorded to obtain the load-depth curves. The hardness (H) and other parameters were determined according to the Oliver–Pharr method as following formulas [14],

$$H = \frac{P}{A} \tag{1}$$

$$A = 24.56 h_c^2 \tag{2}$$

$$h_c = h - \varepsilon \frac{P}{S} \tag{3}$$

$$E_r = \frac{\sqrt{\pi}}{2} \times \frac{S}{\sqrt{A}} \tag{4}$$

$$\frac{1}{E_r} = \frac{1-v^2}{E} + \frac{1-v_i^2}{E_i} \tag{5}$$

where P is the peak indentation load; A is the project area of the hardness impression; h_c is the contact depth deduced from the resultant load-displacement curves; h is the maximum depth of penetration; ε is the indenter geometry constant. For the conical indenter, ε has an empirical value of 0.75; S is the stiffness determined by the upper portion of the unloading data; E_r is the reduced modulus; E and v are the elastic properties of the samples, and Young's modulus and Poisson's ratio of the diamond indenter are E_i = 1440 GPa and v_i = 0.07 [15].

2.2.3. Vickers Indentation Test

A Vickers diamond indenter (AKASHI, AVK-A, Tokyo Japan) was used for measuring the hardness and fracture toughness of ceramics with a polished surface. The indentation load of 19.6 N was applied and the holding time was 15 s. The hardness and fracture toughness were determined by the indentation fracture technique and the geometry patterns of the indentation and cracks were observed by SEM. The values of hardness (H) was measure according to the ASTMC 1327-99 and fracture toughness (K_{IC}) was proposed by Anstis [16], which were calculated by following formulas,

$$H = 1.8544 \frac{P}{d^2} \tag{6}$$

$$K_{IC} = 0.016 \left(\frac{E}{H}\right)^{1/2} \left(\frac{P}{C^{3/2}}\right) \tag{7}$$

where, the value of P is 19.6 N, d is the average length of diagonal under indentation, E is the Young's modulus obtained from the nanoindentation test and C is the length of crack measured from the center of the indentation.

3. Result and Discussion

Figure 1 shows the XRD patterns of the BTWC ceramics sintered at different temperatures. The diffraction peaks marked by rhombs were well-indexed as $Bi_4Ti_3O_{12}$ with an orthorhombic structure and a space group of B2cb (41) in the light of JCPDS card # 72-1019. It suggests that all WO_3 and Cr_2O_3 doped were successfully diffused into the crystal lattice of $Bi_4Ti_3O_{12}$, forming a solid solution with the matrix. Considering the preparing process of BTWC ceramics, WO_3 of 5 mol % was firstly added into the raw material according to the stoichiometric composition of $Bi_4Ti_{2.95}W_{0.05}O_{12.05}$ (BITW), thus these W^{6+} introduced will occupy those Ti^{4+} vacancies designed in the starting composition of powders. While Cr_2O_3 of 0.2 wt % as a fully redundant component were then added into the as-calcined BITW powders, thus Cr^{3+} can only substitute Ti^{4+} in the later sintering process of ceramics [17]. In the [TiO_6] octahedron of $Bi_4Ti_3O_{12}$, the substitution of W^{6+} and Cr^{3+} for Ti^{4+} could be attributed to their similar ionic radius (W^{6+}: 0.600 Å, Cr^{3+}: 0.615 Å and Ti^{4+}: 0.605 Å) and matching coordination number based on the theory of crystal chemistry. However, a secondary phase (asterisked) was detected in all samples, which was identified as $Bi_2Ti_2O_7$ with a cubic structure according to JCPDS card # 32-0118. In the preparation process of BTWC ceramics, $Bi_2Ti_2O_7$ is prone to form in case of an initial Ti/Bi ratio higher than 3/4 during calcining and may also be resulted from the decomposition of $Bi_4Ti_3O_{12}$ during sintering [18]. According to the intensity ratio between the impurity phase and the total phases [19], the phase content of $Bi_2Ti_2O_7$ in each sample can be figured out as follows: 6.63% (1050 °C), 2.52% (1075 °C), 2.46% (1100 °C), 10.18% (1125 °C) and 3.27% (1150 °C), respectively. The sample sintered at 1125 °C seems to contain more $Bi_2Ti_2O_7$ than the others, which may be caused by the heavy decomposition of $Bi_4Ti_3O_{12}$ at this sintering temperature.

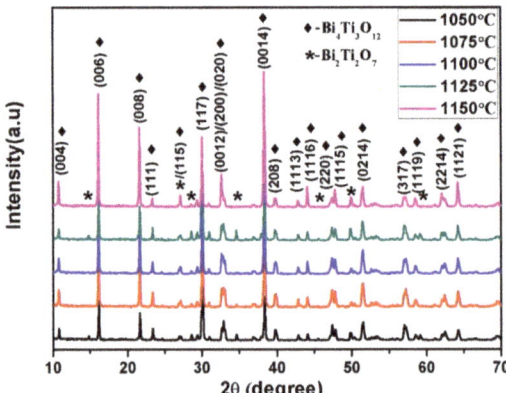

Figure 1. wt %XRD patterns of BTWC ceramics sintered at different temperatures.

Figure 2 shows the microstructures on the natural surfaces of BTWC ceramics sintered at different temperatures. It can be found that all these samples were mainly composed of the plate-like grains with random orientation. This special grain morphology with a high aspect ratio is contributed by a higher grain growth rate along the a-b plane of the crystal [9], which is essentially related to a lower interfacial energy in this crystallographic plane [17]. With the increase of sintering temperatures from 1050 to 1150 °C, the average length of plate-like grains slowly increases from 3.2, 4.1, 5.3 to 7.6 μm in the prophase, but soars to 16.1 μm in the end. The sample sintered at 1150 °C (Figure 2e) presents an extreme grain growth, which may be attributed to the formation of liquid phase accelerating the grain boundary diffusion at the higher sintering temperature. In addition, it can be seen from Figure 2a that some pores could be observed in the sample sintered at 1050 °C. The porosity could reflect the densification effect of ceramics, and some appropriate donor dopants are suggested to favor the densification process of $Bi_4Ti_3O_{12}$ ceramics by a solute drag mechanism [20]. Also, the liquid phase occurring in the grain boundary during sintering may also promote the densification process of ceramics [21]. On the other hand, one can also see that some small polyhedral grains were mingled with these big plate-like grains. These heteroid grains are indexed as $Bi_2Ti_2O_7$ impurity, and their apparent amount basically agree with their phase content for each sample. Here, the sample sintered at 1125 °C seems to contain more impurity as shown in Figure 2d and the inserted figure, a normal growth of $Bi_4Ti_3O_{12}$ grains may be depressed by the competitive growth of $Bi_2Ti_2O_7$ grains [22]. The microstructural evolution of BTWC ceramics with the sintering temperatures are summarized in Table 1. The relative density of all the samples exceed 93%, especially, the sample sintered at 1150 °C obtained a high density of 97.83%, which may be profited from both its intrinsic less impurity phase and extrinsic more liquid phase during sintering.

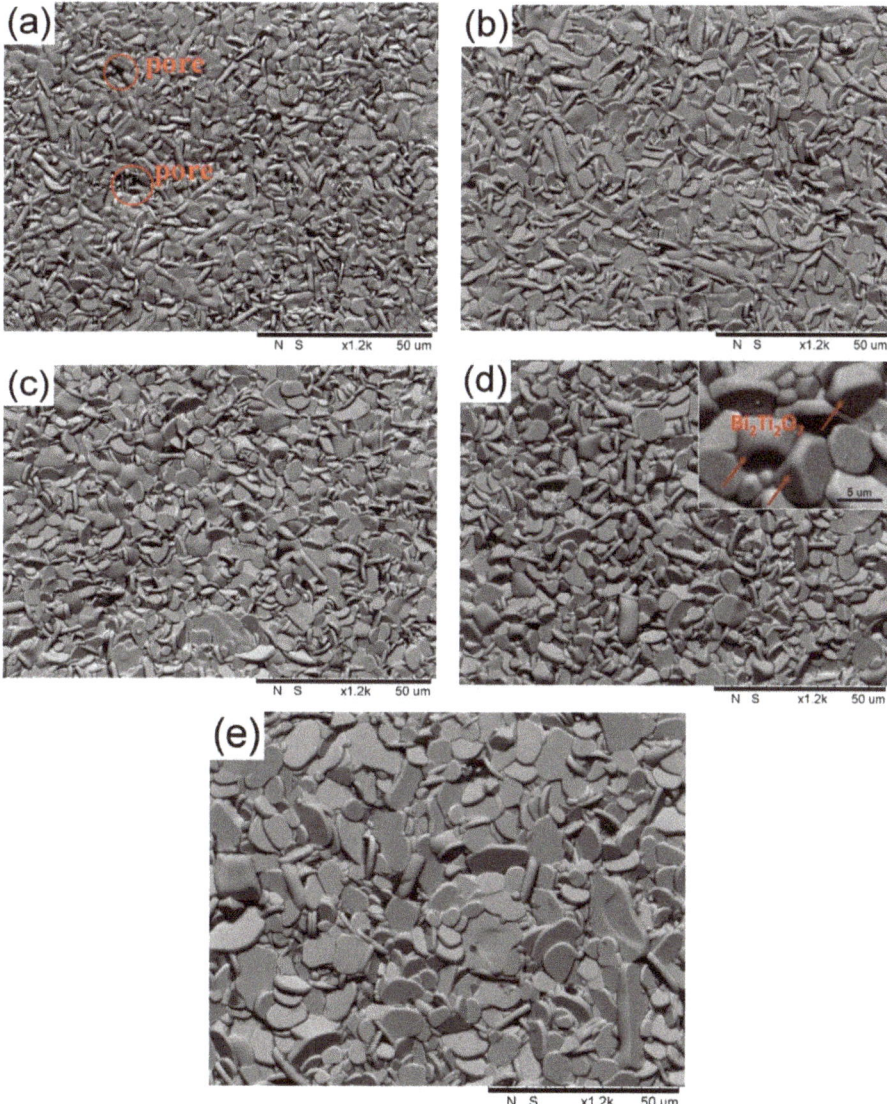

Figure 2. SEM photos on the natural surfaces of BTWC ceramics sintered at different temperatures: (**a**) 1050 °C; (**b**) 1075 °C; (**c**) 1100 °C; (**d**) 1125 °C; (**e**) 1150 °C.

Table 1. Microstructural evolution of BTWC ceramics with the sintering temperatures.

Sintering Temperature	1050 °C	1075 °C	1100 °C	1125 °C	1150 °C
Phase content of $Bi_2Ti_2O_7$ (%)	6.63	2.52	2.46	10.18	3.27
Average grain size of $Bi_4Ti_3O_{12}$ (μm)	3.2	4.3	5.1	7.6	16.1
Relative density of ceramics (%)	95.42	97.09	95.36	96.71	97.83

Figure 3 shows the typical load-displacement curves of BTWC ceramics sintered at different temperatures, which are derived from the nanoindentation test. A peak load of 100 mN was applied to these samples. It can be seen that both the loading curves and the unloading curves were successive and smooth for all samples, that is to say that there is no local brittle fracture occurring in the whole deformation process of BTWC ceramics, thus a fine ductility could be considered for them when subjected to the applied load. In addition, the maximal displacement of samples presents a slight increase following the increase in sintering temperatures, and a significant increase was observed at 1150 °C, which indicates that the change of sintering temperature leads to the evolution of mechanical behavior for BTWC ceramics. In fact, a different elastic–plastic deformation mechanism was concealed under the applied stress. Here, the inserted map in Figure 3 describes the hardness of BTWC ceramics as a function of their sintering temperatures. As can be seen that the hardness value showed a continuous downtrend with the increase of sintering temperatures. This result can be explained by the classical Hall–Petch relation as follows [23],

$$H = H_0 + kd^{-1/2} \tag{8}$$

where, H_0 and k are material constants, d is the grain size of materials. It shows that a higher hardness is usually existing in the fine-grained materials. For BTWC ceramics, it has been identified by SEM that the grain size increase with the increase of sintering temperatures, which determines the decrease of hardness by the Hall–Petch relation. Therefore, the fine-grained ceramics are prone to have a harder plastic deformation because of its higher hardness.

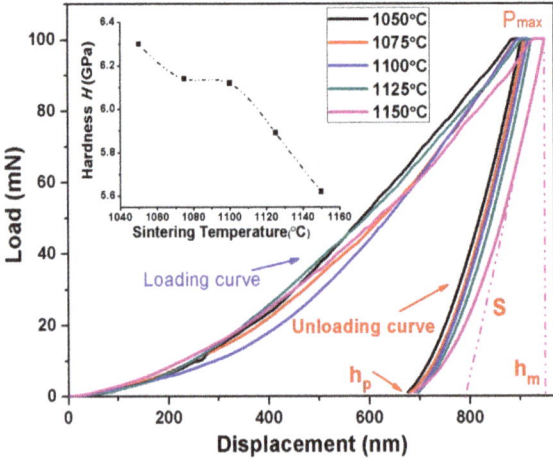

Figure 3. Load-displacement curves of BTWC ceramics sintered at different temperatures in the nanoindentation test.

Here, some mechanical properties of BTWC ceramics obtained by the nanoindentation test are summarized in Table 2. The Young's modulus (E) of solid materials indicates the ability to resist the elastic deformation. As for ceramics, it is considered to have some connection with the strength of ionic bonds and covalent bonds, the porosity and the second grains [24]. The highest Young's modulus of 98.57 GPa is obtained by the sample sintered at 1125 °C, which has the highest phase content of $Bi_2Ti_2O_7$. It has been reported that $Bi_2Ti_2O_7$ has a smaller deformation than $Bi_4Ti_3O_{12}$ when they are subjected to external force, due to the lower axial ratio of its cubic cell [25]. On the other hand, all the maximum depth (h), contact depth (h_c) and residual deformation depth (h_p) roughly present an uptrend with the increase of sintering temperatures. The sample sintered at 1050 °C responds a smaller deformation (both elastic and plastic) to the applied stress compared with the others, which could be ascribed to the fact that the fine-grained materials may bring about additional obstacles for dislocation movement in the adjacent grains. However, the residual depth caused by the plastic deformation is abundant for BTWC ceramics. The proportion of residual deformation depth in the total indentation penetration depth exceeds 50%. Thus, the plastic deformation dominates the total deformation of BTWC ceramics. It suggests that BTWC ceramics have a stronger ability to resist the brittle fracture.

Table 2. Mechanical properties of BTWC ceramics determined by the nanoindentation test.

Sintering Temperature	H (GPa)	E (GPa)	h (nm)	h_c (nm)	h_p (nm)
1050 °C	6.30	96.07	904.51	795.71	644.07
1075 °C	6.14	98.13	911.15	805.44	663.29
1100 °C	6.12	93.49	918.08	807.19	675.80
1125 °C	5.89	98.57	924.91	822.58	662.88
1150 °C	5.62	90.00	951.34	842.80	682.11

It is worth mentioning that the apparent hardness of materials tested directly by the nanoindentation is different with its true hardness, which is due to the indentation size effect resulted from different loads [26,27]. Li and Bradt proposed a proportional specimen resistance (PSR) model as an adequate approach to explain the nanoindentation data of ceramics [28]. In this model, the relationship between the effective indentation load and the indentation dimension can be described by the following formula:

$$\frac{P_{max}}{h_c} = \alpha_1 + \alpha_2 h_c \qquad (9)$$

where, P_{max} is the maximum applied load and h_c is the corresponding indentation contact depth; a_1 and a_2 are two constants of the test material, which is determined by its elastic and plastic properties, respectively. In addition, according to the energy balance consideration proposed by Quinn [29], a_1 and a_2 are affiliated with the energies dissipated in the process of producing a new surface of an unit area and forming the irreversible deformation of an unit volume, respectively. For a nanoindentation test with a Berkovich indenter, both a_1 and a_2 are a measure parameters of the true hardness, H_0, which can be determined directly by the following formulas:

$$H_{01} = \frac{P_{max} - \alpha_1 h_c}{24.5 h_c^2} \qquad (10)$$

$$H_{02} = \frac{\alpha_2}{24.5} \qquad (11)$$

where, H_{01} and H_{02} are the true hardness related to the a_1 and a_2, respectively. a_1 and a_2 can be derived from the plot of P_{max}/h_c versus h_c according to Equation (9). To get the plots of P_{max}/h_c versus h_c, several peak loads from 50, 100, 150 to 200 mN were applied to BTWC ceramics. Here, taking the sample sintered at 1050 °C to show the load-displacement curves at different peak loads (Figure 4). The others have a similar result.

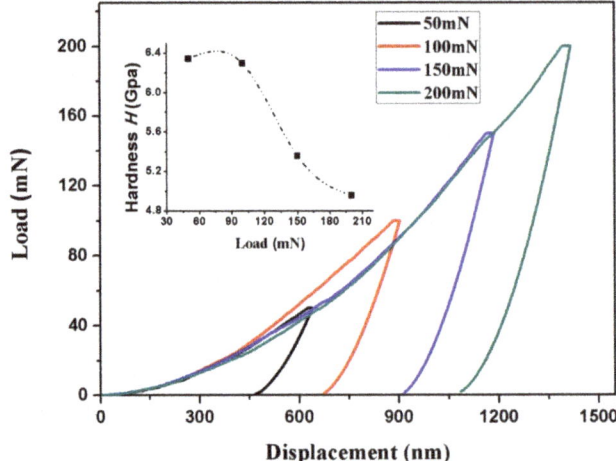

Figure 4. Load-displacement curves at different peak loads for BTWC ceramics sintered at 1050 °C.

As can be seen from Figure 4, both the maximum penetration depth and the residual depth are observed to increase with increasing peak loads. However, the hardness gained by the nanoindentation test presented a downtrend with increasing peak loads as shown in the inserted map. The hardness gains the lowest value of 4.96 GPa at the peak load of 200 mN, while the highest value of 6.34 GPa at 50 mN. It shows that the indentation size effect is significant and the hardness is dependent on the peak load. Based on Equation (9), the P_{max}/h_c versus h_c curve was depicted in Figure 5, and the fitting result (R^2 = 0.95) gives out the value of a_1 and a_2 as 0.045 mN/nm and 9.047 × 10^{-5} nN/nm^2, respectively. And then, the value of true hardness (H_{01} and H_{02}) at different peak loads can be evaluated according to Equations (10) and (11).

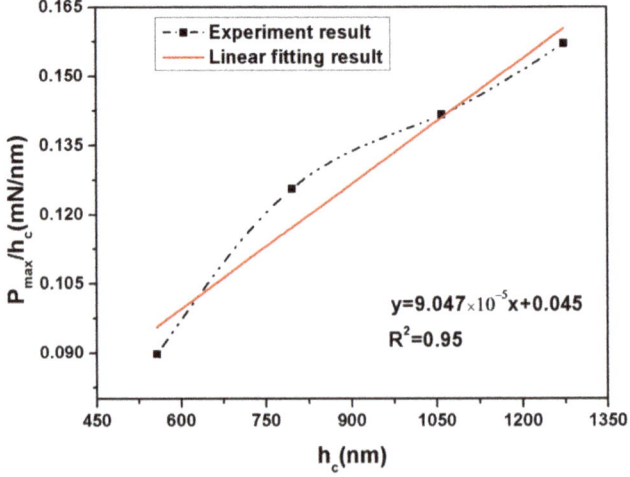

Figure 5. Plots of P_{max}/h_c versus h_c and the linear fitting result.

Based on the PSR model, H_{01} and H_{02} as a function of the peak load are shown in Figure 6. It can be seen that H_{01} fluctuates around H_{02} with increasing peak loads. Their central value are located at

~3.8 GPa, which are much lower than the testing value of hardness (H = 6.3 GPa). This result proves the size effect existing in the nanoindentation test for hardness. In addition, the PSR true hardness of BTWC ceramics sintered at different temperatures were shown in Figure 7.

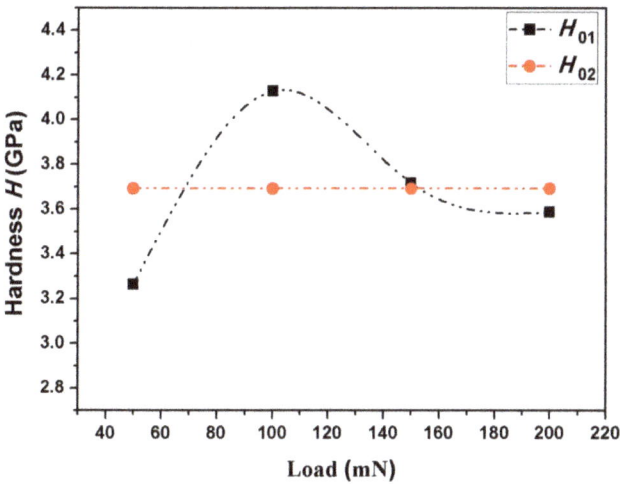

Figure 6. H_{01} and H_{02} of the sample sintered at 1050 °C as a function of peak loads.

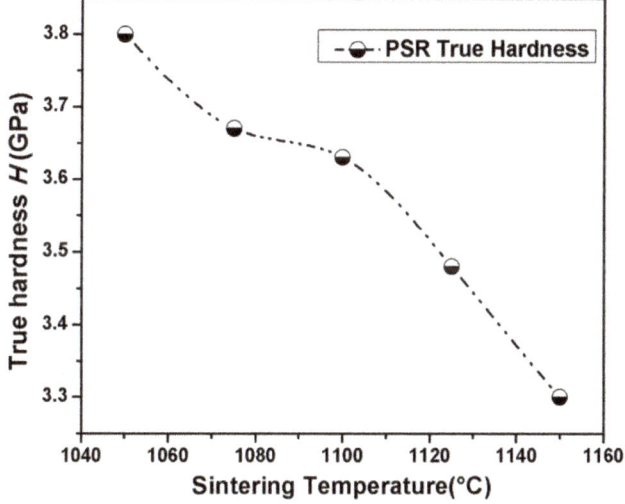

Figure 7. PSR true hardness of BTWC ceramics sintered at different temperatures.

As we know, although the nanoindentation technique fits for investigating the elastic–plastic transition behavior of ceramics at the scale of nanometers, cracks around the indentation can't be observed easily, since a small amount of applied stress is unlikely to produce cracks. Therefore, it is unavailable to test the fracture toughness for ceramics. On the other hand, the Vickers based on a conventional microindentation technique could create both indentations and cracks through a stronger stress field applied for brittle materials. Since such cracks and indentations can be clearly observed by

SEM, this method is usually employed to test the fracture toughness for brittle materials including ceramics and glasses, etc. based on the theory of fracture mechanics.

The typical patterns of the indentation and crack produced by the Vickers are shown for BTWC ceramics in Figure 8. In Figure 8a, it can be seen that a symmetric rhombic indentation was induced by the Vickers diamond indenter on the surface of ceramics, and four cracks straightly propagated along the diagonal direction of rhombic indentation. In the mechanical model of sharp indenter applied for brittle materials, the stress field contributes two superposable components including an elastic (reversible) part and a residual (irreversible) part to the net driving force on the crack system [30]. At the indentation surface, the elastic component is compressive, while the residual component is tensile. Thus the radial cracks grow to their final lengths as the indenter is unloaded, i.e., as the restraining elastic field is removed. Therefore, the area within the rhombic indentation is regarded as the plastic deformation zone formed by the application of sharp indenter, while the crack area outside of the rhombic indentation is considered as the elastic deformation zone used for releasing partial strain energy. Thus the characteristics of cracks can be related to the fracture characteristic of materials. For the sample sintered at 1050 °C, four cracks are observed to linearly extend for a long distance and then gradually disappear in Figure 8a, which is known as a normal model of crack propagation in piezoceramics subjected to a small indentation load. However, this normal crack propagation has changed in the sample sintered at 1125 °C as shown in Figure 8b. Firstly, all cracks are shorter and thinner; secondly, the main crack initiates accompanied with a secondary crack; thirdly, some cracks deflect from its initial propagation direction after extending a short distance, and even a few cracks are tardily split into some slight branches following the main crack. Here, the crack shortening, the secondary crack, the crack defection or the crack branching all them could be attributed to the microstructural aspects (including structural inhomogeneity of matrix, larger aspect ratio of grains and microcracks pre-existing in matrix, etc.) of ceramics according to the description in reference [31]. In addition, it has been identified by SEM and XRD that this sample contains a large number of $Bi_2Ti_2O_7$ second grains, which tend to cause many residual stress in the sintering process of ceramics because of the thermal expansion mismatch between two different phases. Hence, there may be some microcracks induced by the residual stress existing in the matrix, which are considered to initiate second cracks around main cracks by their slow propagation under the stress field. Moreover, they can also weaken the stress field around crack tips (causing the crack defection or branching), or depress the driving force of crack propagation (causing the crack shortening).

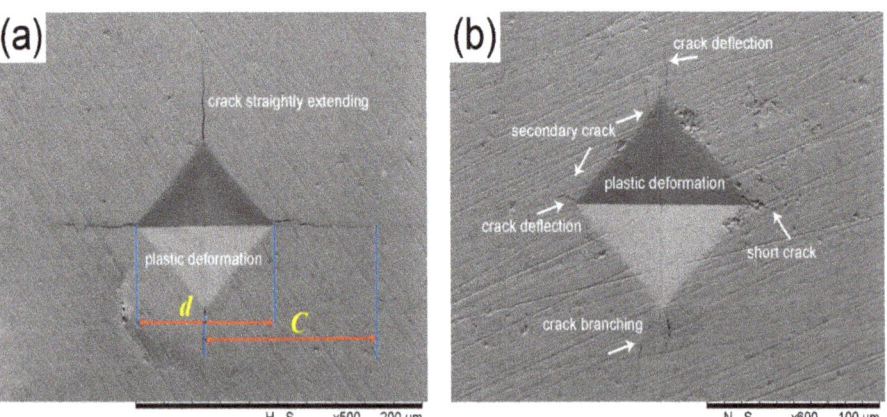

Figure 8. Patterns of Vickers indentation and resulting cracks derived from BTWC ceramic sintered at (**a**) 1050 °C and (**b**) 1125 °C.

Figure 9 displays H and K_{IC} of BTWC ceramics as a function of their sintering temperatures beneath the Vickers diamond indenter, and their values including the average length of diagonal (d) and crack length (C) are shown in Table 3. It can be seen that the hardness decreased gradually with increasing the sintering temperatures from 1050 to 1150 °C, which is just opposite to the uptrend of grain size with the sintering temperatures (Table 1). Therefore, this result also agrees with the Hall-Petch equation which indicates the inverse relation between the hardness and the grain size for brittle ceramics. Moreover, for each sample, its PSR true hardness are basically equal to the Vickers hardness as observed from Figures 7 and 9. In fact, the indentation size effect also exists in Vickers test. Vickers hardness-load curve of brittle ceramics usually presents a hardness-platform with increasing applied loads [29], and the level value of hardness is identified as the true hardness of materials. In this experiment, the indentation load of 19.8 N could be considered to fit for BTWC ceramics in terms of such a fine morphology of cracks shown in Figure 8, and it corresponds to the applied load for the hardness approaching its constant value, which is similar with the load dependence of the PSR true hardness in the nanoindentation test.

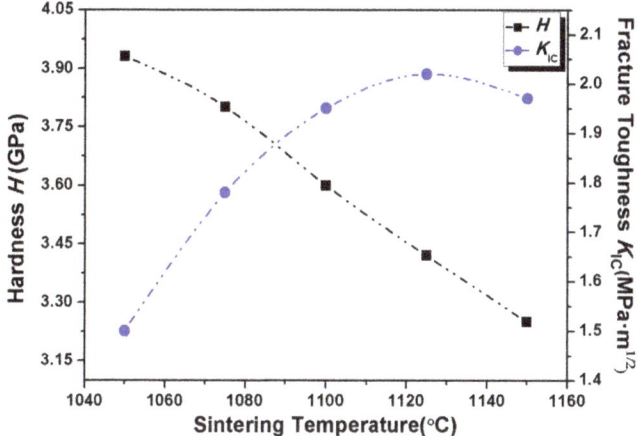

Figure 9. Hardness and fracture toughness of BTWC ceramics sintered at different temperatures.

Table 3. The values of parameters obtained by Vickers indentation test.

Sintering Temperature	d (µm)	H (Gpa)	P (N)	C (µm)	K (MPa·m$^{1/2}$)
1050 °C	96.19	3.93	19.6	102.16	1.50
1075 °C	97.83	3.80	19.6	92.86	1.78
1100 °C	100.59	3.59	19.6	87.69	1.95
1125 °C	103.15	3.42	19.6	88.74	2.02
1150 °C	105.83	3.25	19.6	88.83	1.97

On contrary, the fracture toughness of BTWC ceramics exhibited an approximate uptrend with increasing sintering temperatures. In the case of a higher sintering temperature, the resulting coarse grains can absorb much of the energy of crack propagation with the help of the crack deflection or branching mechanism [21]. As a result, the fracture toughness of materials tends to increase with increase of grain size. However, the highest fracture toughness of 2.02 MPa·m$^{1/2}$ is given to the sample sintered at 1125 °C rather than the sample sintered at 1150 °C, which has the largest grain size. Maybe, we can understand this result based on the fact as follows. Firstly, this sample contains more $Bi_2Ti_2O_7$ than the others. It has been reported that the second phase with a higher thermal expansion coefficient will induce a compressive stress field in the matrix to absorb the fracture energy, blunt crack tips and shield crack propagation leading to a higher fracture toughness [32]. Besides, the second phase also induces microcracks to the matrix, which are considered as one type of toughening mechanism for ceramics. Therefore, the sample sintered at 1125 °C gains a larger fracture toughness than the others, which could be ascribed to its higher sintering temperatureand numerous second phase grains.

4. Conclusions

Indentation behavior and mechanical properties of W/Cr co-doped $Bi_4Ti_3O_{12}$ ceramics (BTWC) sintered at different temperatures were investigated. XRD revealed that $Bi_4Ti_3O_{12}$ as the major phase and $Bi_2Ti_2O_7$ as the second phase coexisted in BTWC ceramics, and the phase content of $Bi_2Ti_2O_7$ varied with sintering temperatures. SEM demonstrated that BTWC ceramics were composed of the plate-like grains with random orientation, and the grain size increased with increase of sintering temperatures. In the nanoindentation test, the hardness value (5.62 GPa~6.3 GPa) of BTWC ceramics was found to decrease with increase of sintering temperatures, which could be explained by the Hall–Petch equation, and their true hardness (3.3 GPa~3.8 GPa) could be calculated by the PSR model considering the indentation size effect. In the Vickers indentation test, these cracks produced tended to be shortened in the sample sintered at a higher temperature (1125 °C), as well as presented deflection and branching. The hardness and the fracture toughness of BTWC ceramics had a contrary variational tendency with sintering temperatures.

Acknowledgments: This work was supported by the Applied Basic Research Program from Sichuan Province (2017JY0091), National Natural Science Foundation of China (Grant No. 11572057 and No. 11702037), China Postdoctoral Science Foundation Funded Project (2017M623025) and Program for Changjiang Scholars and Innovative Research Team (IRT14R37).

Author Contributions: Q. Wang and Y.Chen conceived and designed the experiments; S.X.Xie, J.G.Xu, Z.Tan and R.Nie performed the experiments; S.X.Xie, Z.Tan, R.Nie and J.G.Zhu analyzed the data; Q.Y.Wang and J.G.Zhu contributed reagents and analysis tools; S.X.Xie and Y.Chen wrote the paper, and all the authors revised the paper.

Conflicts of Interest: The authors declare no conflicts of interest.

References

1. Meetham, G.W. High-temperature materials—A general review. *J. Mater. Sci.* **1991**, *26*, 853–860. [CrossRef]
2. Yan, H.; Zhang, H.; Ubic, R.; Reece, M.J.; Liu, J.; Shen, Z.; Zhang, Z. A Lead-Free High-Curie-Point Ferroelectric Ceramic, $CaBi_2Nb_2O_9$. *Adv. Mater.* **2005**, *17*, 1261–1265. [CrossRef]
3. Damjanovic, D. Materials for high temperature piezoelectric transducers. *Curr. Oper. Solid State Mater. Sci.* **1998**, *3*, 469–473. [CrossRef]
4. Zhu, X.; Xu, J.; Meng, Z.; Zhu, J.; Zhou, S.; Li, Q.; Liu, Z.; Ming, N. Microdisplacement characteristics and microstructures of functionally graded piezoelectric ceramic actuator. *Mater. Des.* **2000**, *21*, 561–566. [CrossRef]
5. Shen, Z.Y.; Li, J.F.; Chen, R.; Zhou, Q.; Shung, K.K. Microscale 1-3-Type (Na,K)NbO_3-Based Pb-Free Piezocomposites for High-Frequency Ultrasonic Transducer Applications. *J. Am. Ceram. Soc.* **2011**, *94*, 1346–1349. [CrossRef] [PubMed]
6. Song, K.; Xu, Y.; Zhao, N.; Zhong, L.; Shang, Z.; Shen, L.; Wang, J. Evaluation of Fracture Toughness of Tantalum Carbide Ceramic Layer: A Vickers Indentation Method. *J. Mater. Eng. Perform.* **2016**, *25*, 3057–3064. [CrossRef]
7. Sebastiani, M.; Johanns, K.E.; Herbert, E.G.; Pharr, G.M. Measurement of fracture toughness by nanoindentation methods: Recent advances and future challenges. *Curr. Oper. Solid State Mater. Sci.* **2015**, *19*, 324–333. [CrossRef]
8. Du, H.; Tang, F.; Luo, F.; Zhu, D.; Qu, S.; Pei, Z.; Zhou, W. Influence of sintering temperature on piezoelectric properties of ($K_{0.5}Na_{0.5}$) NbO_3–$LiNbO_3$ lead-free piezoelectric ceramics. *Mater. Res. Bull.* **2007**, *42*, 1594–1601. [CrossRef]
9. Yan, H.; Li, C.; Zhou, J.; Zhu, W.; He, L.; Song, Y.; Yu, Y. Influence of sintering temperature on the properties of high Tc bismuth layer structure ceramics. *Mater. Sci. Eng. B* **2002**, *88*, 62–67. [CrossRef]
10. Khokhar, A.; Mahesh, M.L.V.; James, A.R.; Goyal, P.K.; Sreenivas, K. Sintering characteristics and electrical properties of $BaBi_4Ti_4O_{15}$ ferroelectric ceramics. *J. Alloys Compd.* **2013**, *581*, 150–159. [CrossRef]

11. Liu, L.; Fan, H.; Ke, S.; Chen, X. Effect of sintering temperature on the structure and properties of cerium-doped 0.94 $(Bi_{0.5}Na_{0.5})$ TiO_3–$0.06BaTiO_3$ piezoelectric ceramics. *J. Alloys Compd.* **2008**, *458*, 504–508. [CrossRef]
12. Hou, J.; Qu, Y.; Vaish, R.; Varma, K.B.R.; Krsmanovic, D.; Kumar, R.V. Crystallographic Evolution, Dielectric, and Piezoelectric Properties of $Bi_4Ti_3O_{12}$:W/Cr Ceramics. *J. Am. Ceram. Soc.* **2010**, *93*, 1414–1421. [CrossRef]
13. Chen, Y.; Pen, Z.; Wang, Q.; Zhu, J. Crystalline structure, ferroelectric properties, and electrical conduction characteristics of W/Cr co-doped $Bi_4Ti_3O_{12}$ ceramics. *J. Alloys Compd.* **2014**, *612*, 120–125. [CrossRef]
14. Oliver, W.C.; Pharr, G.M. An improved technique for determining hardness and elastic modulus using load and displacement sensing indentation experiments. *J. Mater. Res.* **1992**, *7*, 1564–1583. [CrossRef]
15. Donnelly, E.; Baker, S.P.; Boskey, A.L.; van der Meulen, M.C. Effects of surface roughness and maximum load on the mechanical properties of cancellous bone measured by nanoindentation. *J. Biomed. Mater. Res. Part A* **2006**, *77*, 426–435. [CrossRef] [PubMed]
16. Lawn, B.R.; Evans, A.G.; Marshall, D.B. Elastic/Plastic Indentation Damage in Ceramics: The Median/Radial Crack System. *J. Am. Ceram. Soc.* **1980**, *63*, 574–581. [CrossRef]
17. Chen, Y.; Xie, S.; Wang, Q.; Zhu, J. Influence of Cr_2O_3 additive and sintering temperature on the structural characteristics and piezoelectric properties of $Bi_4Ti_{2.95}W_{0.05}O_{12.05}$ Aurivillius ceramics. *Prog. Nat. Sci. Mater. Int.* **2016**, *26*, 572–578. [CrossRef]
18. Su, W.F.; Lu, Y.T. Synthesis, phase transformation and dielectric properties of sol-gel derived $Bi_2Ti_2O_7$ ceramics. *Mater. Chem. Phys.* **2003**, *80*, 632–637. [CrossRef]
19. Liu, J.; Bai, W.; Yang, J.; Xu, W.; Zhang, Y.; Lin, T.; Meng, X.; Duan, C.G.; Tang, X.; Chu, J. The Cr-substitution concentration dependence of the structural, electric and magnetic behaviors for Aurivillius $Bi_5Ti_3FeO_{15}$ multiferroic ceramics. *J. Appl. Phys.* **2013**, *114*, 234101. [CrossRef]
20. Villegas, M.; Caballero, A.C.; Fernandez, J.F. Modulation of Electrical Conductivity Through Microstructural Control in $Bi_4Ti_3O_{12}$-Based Piezoelectric Ceramics. *Ferroelectrics* **2002**, *267*, 165–173. [CrossRef]
21. Gu, M.; Xu, H.; Zhang, J.; Wei, Z.; Xu, A. Influence of hot pressing sintering temperature and time on microstructure and mechanical properties of TiB_2/TiN tool material. *Mater. Sci. Eng. A* **2012**, *545*, 1–5. [CrossRef]
22. Villegas, M.; Jardiel, T.; Caballero, A.C.; Fernández, J.F. Electrical Properties of Bismuth Titanate Based Ceramics with Secondary Phases. *J. Electroceramics* **2004**, *13*, 543–548. [CrossRef]
23. Pande, C.S.; Cooper, K.P. Nanomechanics of Hall-Petch relationship in nanocrystalline materials. *Prog. Mater. Sci.* **2009**, *54*, 689–706. [CrossRef]
24. Chen, Y.; Miao, C.; Xie, S.; Xu, L.; Wang, Q.; Zhu, J.; Guan, Z. Microstructural evolutions, elastic properties and mechanical behaviors of W/Cr Co-doped $Bi_4Ti_3O_{12}$ ceramics. *Mater. Des.* **2016**, *90*, 628–634. [CrossRef]
25. Fisher, E.S.; Manghnani, M.H. Effect of Axial Ratio Changes on the Elastic Moduli and Grüneisen γ for Lower Symmetry Crystals. *J. Appl. Phys.* **1970**, *41*, 5059–5062. [CrossRef]
26. Peng, Z.; Gong, J.; Miao, H. On the description of indentation size effect in hardness testing for ceramics: Analysis of the nanoindentation data. *J. Eur. Ceram. Soc.* **2004**, *24*, 2193–2201. [CrossRef]
27. Wang, H.; Huang, Z.; Lu, Z.; Wang, Q.; Jiang, J. Determination of the elastic and plastic deformation behaviors of Yb:$Y_3Al_5O_{12}$ transparent ceramic by nanoindentation. *J. Alloys Compd.* **2016**, *682* (Suppl. C), 35–41. [CrossRef]
28. Li, H.; Bradt, R.C. The microhardness indentation load/size effect in rutile and cassiterite single crystals. *J. Mater. Res.* **1993**, *28*, 917–926. [CrossRef]
29. Quinn, J.B.; Quinn, G.D. Indentation brittleness of ceramics: A fresh approach. *J. Mater. Res.* **1997**, *32*, 4331–4346.
30. Anstis, G.R.; Chantikul, P.; Lawn, B.R.; Marshall, D.B. A Critical Evaluation of Indentation Techniques for Measuring Fracture Toughness: I, Direct Crack Measurements. *J. Am. Ceram. Soc.* **1981**, *64*, 533–538. [CrossRef]
31. Wu, C.C.; Freiman, S.W.; Rice, R.W.; Mecholsky, J.J. Microstructural aspects of crack propagation in ceramics. *J. Mater. Sci.* **1978**, *13*, 2659–2670. [CrossRef]
32. Cutler, R.A.; Virkar, A.V. The effect of binder thickness and residual stresses on the fracture toughness of cemented carbides. *J. Mater. Sci.* **1985**, *20*, 3557–3573. [CrossRef]

© 2018 by the authors. Licensee MDPI, Basel, Switzerland. This article is an open access article distributed under the terms and conditions of the Creative Commons Attribution (CC BY) license (http://creativecommons.org/licenses/by/4.0/).

Article

Study of the Fatigue Crack Growth in Long-Term Operated Mild Steel under Mixed-Mode (I + II, I + III) Loading Conditions

Grzegorz Lesiuk [1],*, Michał Smolnicki [1], Dariusz Rozumek [2], Halyna Krechkovska [3], Oleksandra Student [3], José Correia [4],*, Rafał Mech [1] and Abílio De Jesus [4]

1. Department of Mechanics, Material Science and Engineering, Wroclaw University of Science and Technology, Smoluchowskiego 25 Wrocław, PL-50372 Wrocław, Poland; michal.smolnicki@pwr.edu.pl (M.S.); rafal.mech@pwr.edu.pl (R.M.)
2. Department of Mechanics and Machine Design, Opole University of Technology, Mikołajczyka 5, PL-45271 Opole, Poland; d.rozumek@po.edu.pl
3. Karpenko Physico-Mechanical Institute of the National Academy of Sciences of Ukraine, Naukova 5, 79060 Lviv, Ukraine; galyna@ipm.lviv.ua (H.K.); student@ipm.lviv.ua (O.S.)
4. INEGI, Faculty of Engineering, University of Porto, Rua Dr. Roberto Frias, Campus FEUP, 4200-465 Porto, Portugal; ajesus@fe.up.pt
* Correspondence: Grzegorz.Lesiuk@pwr.edu.pl (G.L.); jacorreia@inegi.up.pt (J.C.)

Received: 4 December 2019; Accepted: 26 December 2019; Published: 1 January 2020

Abstract: The paper presents an analysis of mixed-mode fatigue crack growth in bridge steel after 100-years operating time. Experiments were carried out under mode I + II configuration on Compact Tension Shear (CTS) specimens and mode I + III on rectangular specimens with lateral stress concentrator under bending and torsion loading type. Due to the lack of accurate Stress Intensity Factor (SIF) solutions, the crack path was modelled with the finite element method according to its experimental observation. As a result, the Kinetic Fatigue Fracture Diagrams (KFFD) were constructed. Due to the change in the tendency of higher fatigue crack growth rates from K_I towards K_{III} dominance for the samples subjected to bending and torsion, it was decided to analyze this phenomenon in detail using electron-scanning microscopy. The fractographic analysis was carried out for specimens subjected to I + III crack loading mode. The mechanism of crack growth in old bridge steel at complex loads was determined and analyzed.

Keywords: mixed-mode fracture; fatigue crack growth; crack growth rate; finite element analysis; crack paths; crack closure; fractography

1. Introduction

Fatigue fracture is one of the most frequent failure reasons for bridge structures. It is worth noting that the research guidelines [1] for old bridge structures are conservative as regard fatigue, namely as concerns the sub-critical period of fatigue crack growth. Of course, this is motivated by the absence of reliable data from old bridge materials, especially original data. The maintenance of historic bridges is still an important issue, especially justified by the fact that every year, the number of structures with more than 100 years of service life grows. It is generally assumed that until the end of the 19th century, the construction material was the puddle iron. Many papers have been devoted to the puddle iron [2–6] in the field of material characterization, including fatigue testing. While state of the art and test procedures are clear for uniaxial loads, despite the diversity of results, there is little work on the multi-axial fatigue of this type of material. Papers [7–9] of the authors' team devoted to the mixed-mode (I + II) fatigue fracture of puddle iron, show significant discrepancies in the expected fatigue-fracture behavior of puddle iron compared with existing mixed-mode criteria. On the other hand, realistic assessment of the structural integrity using modern tools such as the finite element method [10–12] requires the substitution of growth models for uniaxial states with models for mixed fracture growth modes. The development of metallurgy at the beginning of the 20th century has enabled the manufacture of homogenous materials with parameters similar to modern low carbon structural steels. However, many papers [13–15] reported the deterioration of mechanical properties caused by the phenomenon of microstructural degradation. In the case of old bridge steels, the same material properties as for modern steels are often assumed. Similarly to modern structural steels, the fatigue crack propagation behavior studies have been limited to uniaxial state analyzes. This paper is intended to contribute to fill the existing gap in literature analyzing the fatigue crack growth process for typical fatigue crack load schemes, i.e., tension and shear (I + II), occurring at the riveted crack initiation points [16], and bending and torsion (I + III), occurring at the bridge structural members [17]. Due to the lack of available experimental data for objects erected at the beginning of the 20th century, it was decided to focus the research on material samples from a structural part extracted from a repaired existing steel bridge (1899–1902) located in Poland [18].

2. Materials and Methods

2.1. Material Data

For mixed-mode fatigue crack growth studies, old mild steel extracted from the old railway bridge is considered. The mentioned bridge is part of railway line 143 (Kalety–Wrocław Mikołajów) and is located in Kluczbork, Poland. The chemical composition analysis was carried out in order to identify the tested steel. Chemical composition of this steel, as well as reference values of typical puddle iron and old mild steel are presented in Table 1. Presented data suggested that the steel from Kluczbork Bridge belongs to an old mild steel group. Before mechanical testing, the metallographic study was performed for the analyzed material. The microstructure of the tested steel is shown in Figure 1. A ferrite grain structure with non-metallic inclusions is observed, corresponding well with carbon content. Static tensile testing was also performed and reported in [18]. A representative true stress–strain curve is shown in Figure 2, from which is reported the following mechanical properties [18]: Young modulus, $E = 212$ GPa; ultimate tensile strength, $\sigma_u = 416$ MPa; yield tensile strength, $Re = 304$ MPa.

Table 1. Chemical composition of the investigated steel and typical values for puddle iron and old mild steel [18].

Materials	C [%]	Mn [%]	Si [%]	P [%]	S [%]
Investigated steel	0.1	0.52	0.0004	0.028	0.03
Typical values for puddle iron	<0.8	0.4	n/a	<0.6	<0.04
Typical values for old mild steel	<0.15	0.2–0.5	variable	<0.06	< 0.15

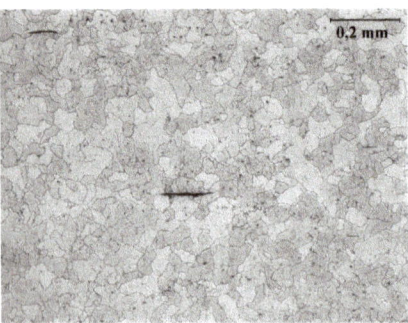

Figure 1. The microstructure of the tested long-term operated mild steel.

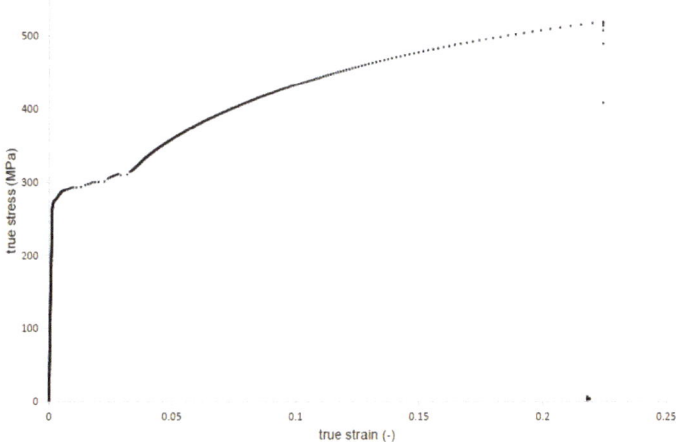

Figure 2. Representative room temperature tensile true stress-true strain curve obtained from the uniaxial tensile test for the old mild steel.

2.2. Mixed-Mode Fatigue-Fracture Characterization—Experimental Details

In order to assess the fatigue-fracture behavior under mixed-mode loading conditions (I + II, I + III), two kinds of experiments were performed in order to determine the fatigue crack propagation rates. For the experimental campaign, two types of specimen were designed (Figure 3). In Figure 3a, the main dimensions of Compact Tension Shear (CTS) specimens according to the original concept of Richard [19] are shown. For mode I + III, the prismatic specimen with the lateral notch was designed and prepared, see Figure 3b.

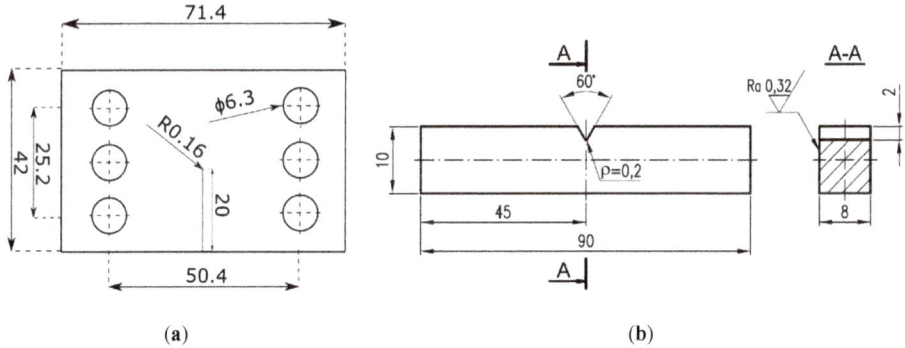

Figure 3. Geometry and dimensions (in mm) of specimens used in both the experimental and numerical campaign: (a) CTS (Compact Tension Shear) specimen (thickness—8 mm); (b) specimen subjected for mode I + III testing.

2.2.1. Mode I + II Test

CTS specimens used in this study presented an initial notch with dimensions: Length, $l = 20$ mm and root radius, $\rho = 0.16$ mm. Detailed geometry of the used specimen is presented in Figure 3a. These notches were cut using electro-discharged machining (EDM).

The main advantage of the CTS specimen is the fact that the mixed-mode loading condition can be performed using a uniaxial hydraulic pulsator. The experimental setup is presented in Figure 4. An unique clevis and gripping system are necessary for testing the CTS specimen. Due to changing the θ angle (defined in Figure 4a), it is possible to generate pure mode I experiment (θ = 0°) as well as "pure shear" state (θ = 90°). Between θ = 0° and θ = 90° mixed-mode loading conditions are generated. All specimens were pre-cracked under pure mode I condition and preserving the linear elastic fracture mechanics conditions. After pre-cracking, the proper mixed-mode experiments were performed for load angle θ = 30° and θ = 45°, respectively. Experimental setup (see Figure 4b) consists of a 100 kN load cell (1), clevis (2), CTS specimen holder (3), light source with polarized filters (4), DinoLite Microscope (5) as well as integrated measurement system operated by computer with MTS FlexTest controller software (6). Tests were conducted with a frequency, $f = 10$ Hz, constant force amplitude, $F = 10$ kN, and load ratio, $R = 0.1$. During the experiment, an additional imaging system was involved in order to periodically capture crack tip point images.

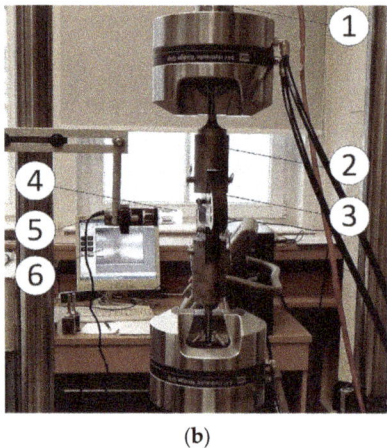

(a) (b)

Figure 4. Experimental mixed-mode I + II loading test: (**a**) Definition of load angle, θ; (**b**) measurement stand: 1—load cell, 2—gripping system, 3—CTS clevis, 4 – light source, 5—digital microscope, 6—imaging system.

2.2.2. Mode I + III Test

The specimens subjected to mixed mode (I + III) condition showed an external, unilateral notch, which was $a_0 = 2$ mm deep and its root radius was, $\rho = 0.2$ mm. The notches in specimens were cut with a cutter, and their surfaces were polished with progressively finer emery papers. The tests combining bending with torsion (mode I + III) were performed on the fatigue test stand MZGS–100 [20–22], enabling the realization of cyclically variable and static (mean) loading. The fatigue test stands MZGS–100 (Figure 5, Opole University of Technology, Opole, Poland) consists of power, control, and loading units. Cyclic loading is obtained by vertical movements of the lever (2) motion in the vertical plane, generated by the inertial force of the rotating disk (1) mounted on flat springs (5). A spring-loaded actuator (3) is fixed to the fatigue testing machine base. The tests were conducted under controlled force (the amplitude of bending moment was controlled) with the frequency of 28.4 Hz. The tests were performed at a constant moment amplitude $M_a = 17.19$ N·m and stress ratio, $R = 0$. The specimen during a test is shown in Figure 5b. The crack growth was observed by an optical method on specimen lateral surfaces with a magnification of 25 times, by recording the number of load cycles through the control box (6). Proportional bending with torsion, for $\alpha = 30°$ and $45°$, was obtained after proper head rotation (4).

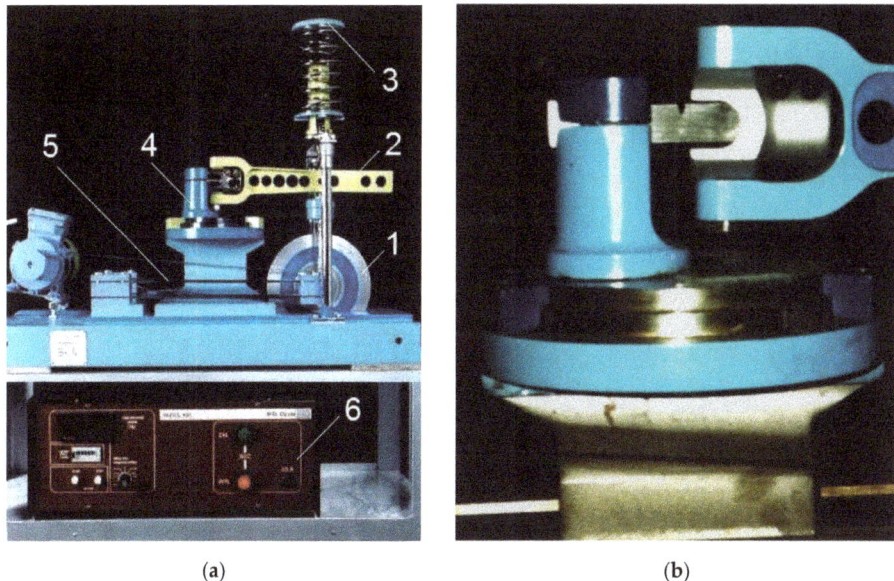

Figure 5. Fatigue test stand MZGS-100: (**a**) Overview; (**b**) specimen during the test.

2.3. Stress Intensity Factors Calculations

In order to evaluate experimental results, stress intensity factors are needed (to calculate values of $\Delta K_{I,II,III}$). In literature, some formulas can be found for CTS specimen cases (Equations (1) and (2)) [19,23]. Nevertheless, these formulas are only valid for initial conditions, i.e., when the crack tip is in the symmetry plane of specimen:

$$K_I = \frac{F \cdot \sqrt{\pi a_0} \cdot \cos\theta}{Wt\left(1 - \frac{a_0}{W}\right)} \sqrt{\frac{0.26 + 2.65\left(\frac{a_0}{W-a_0}\right)}{1 + 0.55\left(\frac{a_0}{W-a_0}\right) - 0.08\left(\frac{a_0}{W-a_0}\right)^2}} \quad (1)$$

$$K_{II} = \frac{F \cdot \sqrt{\pi a_0} \cdot \sin\theta}{Wt\left(1 - \frac{a_0}{W}\right)} \sqrt{\frac{-0.23 + 1.4\left(\frac{a_0}{W-a}\right)}{1 - 0.67\left(\frac{a_0}{W-a_0}\right) - 2.08\left(\frac{a_0}{W-a_0}\right)^2}} \quad (2)$$

where: F—applied force, a_0—initial crack length (notch + precrack), θ—loading angle, W—specimen width.

Analogous formulas for specimens prepared for mode I and III tests are presented in Equations (3) and (4) [21]:

$$K_I = Y_I \sigma \cos^2\alpha \sqrt{\pi(a_0 + a)} \quad (3)$$

$$K_{III} = Y_{III} \sigma \cos\alpha \sin\alpha \sqrt{\pi(a_0 + a)} \quad (4)$$

In Equations (3) and (4), σ represents stress level, α the loading angle, and Y_I and Y_{III} are dimensionless geometric factors defined as [24,25]:

$$Y_I = \frac{5}{\sqrt{20 - 13\left(\frac{a_0 + a}{w}\right) - 7\left(\frac{a_0 + a}{w}\right)^2}} \quad (5)$$

$$Y_{III} = \sqrt{\frac{2w}{a_0 + a}} \tan \frac{\pi(a_0 + a)}{2w} \quad (6)$$

Due to the limitations of previous formulae, the finite element method (FEM) was applied to support the experimental investigations. Abaqus FEM code (version 6.13-2) was used in all cases. The two-dimensional specimen was modelled using a geometry like the one shown in Figures 6–8. For the CTS specimen, boundary conditions used in the analysis are presented in Figure 6a. Force components were defined to show a resultant force in the desired direction and a null resultant moment. Forces satisfying these conditions were calculated as presented in Equations (7)–(9). Pins were modelled as kinematic couplings, but rotation about the z-axis was allowed to prevent over stiffness. Discretization was made mostly using CPS4R elements (4-noded bilinear plane stress quadrilateral elements with reduced integration and hourglass control). Near the crack tip, triangular elements were used to ensure that contours will be of circular shape. The Contour Integral Method requires that all elements are of the same type, so triangular elements were automatically converted from quadrilaterals with one side collapsed. The finite element mesh is presented in Figure 6b. Stress intensity factors were calculated using the Contour Integral Method. To handle singularity problem, nodes at the crack tip were translated by 0.25 times of the element length. The whole process of determining stress intensity factors during crack growth consisted of simulation for every observed crack growth increment. During the simulation, the re-meshing approach has been adopted. However, it is worthwhile to underline that the presented approach is sensitive to a length scale. Therefore, the authors selected this crack growth increment that allows predicting results in the same manner as in the experimental campaign. For every simulation, the direction of fatigue crack growth and crack growth increment corresponds with the real crack paths registered during the experiments.

A similar approach was successfully implemented in previous authors' papers devoted to mixed mode fatigue crack growth in 19th century puddle iron [7,8]. For modern constructional steels [21], the MTS (Maximum Tangential Stress) criterion works well enough, and it allows for the prediction of the fatigue crack path in the numerical environment, but in case of the long-term operated metallic material, it can lead to errors [7,8]. On the other hand, for the long-term operated materials, it is more reasonable to implement real crack paths (and crack growth increments) from the experiments. This approach was selected because the criterion used in Abaqus (version 6.13-2) allows for the prediction of the crack growth direction using conventional SED (Strain Energy Density) or MTS criteria. In the presented case, the MTS criterion does not predict well enough fatigue crack growth direction—see results in Section 3.1. This procedure was automated using an Abaqus script (version 6.13-2) written in Python (version 2.6.2). SIFs obtained in that way are in reasonable conformity with analytical formulas (Equations (1) and (2)) for the cases where these formulas are valid. Figure 7 illustrates the Huber–Mises stress distribution in a CTS specimen loaded at a load angle, $\theta = 45°$. The same figure also represents a geometric model defined for a real crack measurement. This geometric model is afterwards used to compute the the SIF using the described numerical model.

$$F_1 = F \cdot \left(0.5 \cos \alpha + \frac{c}{b} \sin \alpha\right) \quad (7)$$

$$F_2 = F \cdot \sin \alpha \quad (8)$$

$$F_3 = F \cdot \left(0.5 \cos \alpha - \frac{c}{b} \sin \alpha\right) \quad (9)$$

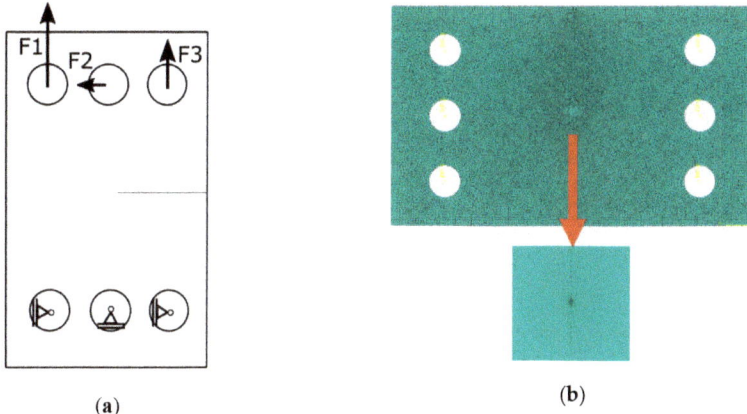

Figure 6. CTS specimen numerical analysis: (**a**) Boundary conditions; (**b**) finite element mesh with detail overview near the crack tip.

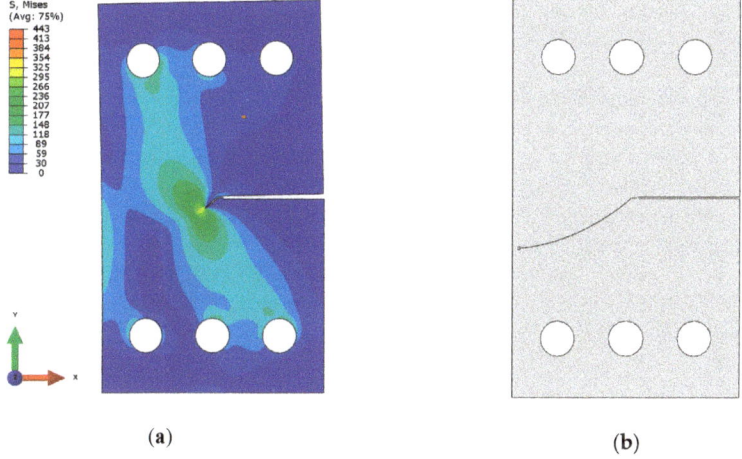

Figure 7. (**a**) Huber–Mises stress distribution in CTS specimen (load angle θ = 45°) with 3 mm long mixed-mode fatigue crack length; (**b**) predicted (posterior—based on the real crack path) crack trajectory in the discrete model used for stress intensity factors calculation.

Numerical simulations using the Finite Element Method for the mixed-mode I + III condition were also performed based on a solid model. The discrete model of the specimen is shown in Figure 8a. The mesh is coarse in the gripping part and more refined in the notch volume. The boundary conditions are established by fixing one side of the specimen and applying concentrated load to imitate bending and torsion on the other side into a reference point. This reference point was then connected to the specimen by a kinematic coupling.

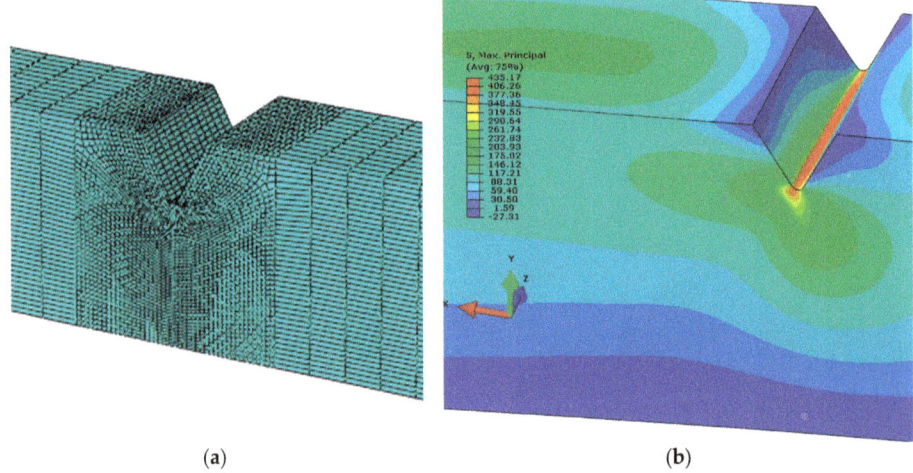

Figure 8. Rectangular specimen subjected to mode I + III testing: (**a**) Notch mesh density; (**b**) principal stress distribution at the initial stage of the crack growth for load angle, $\alpha = 30°$.

3. Results

In general, the nonlinear crack path can be observed during experiments, and its plastic deformation in the last part of the crack path. Due to the plane constraint of the specimen and linear elastic fracture mechanics (LEFM) limitations, not all crack tip points from real crack paths were considered in all calculations. This posterior crack path was generated step-by-step via python script in Abaqus Simulia (version 6.13-2) environment in order to provide good conformity with a real crack growth trajectory. From the experiment, the a-N curves were collected, see Figure 9. Experimentally determined fatigue crack growth curves were analyzed for $\theta = a = 30°$, and $\theta = a = 45°$. Then, kinetic fatigue-fracture diagrams (KFFD) including crack growth rates, $\frac{da}{dN}$ were constructed as a function of ΔK_i for different loading modes i. The KFFDs are presented in Figures 10 and 11.

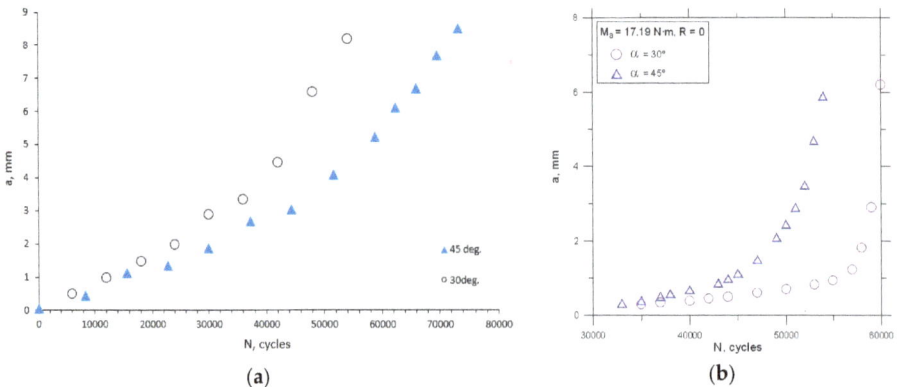

Figure 9. Fatigue crack growth curves for: (**a**) Mode I and II; (**b**) mode I and III.

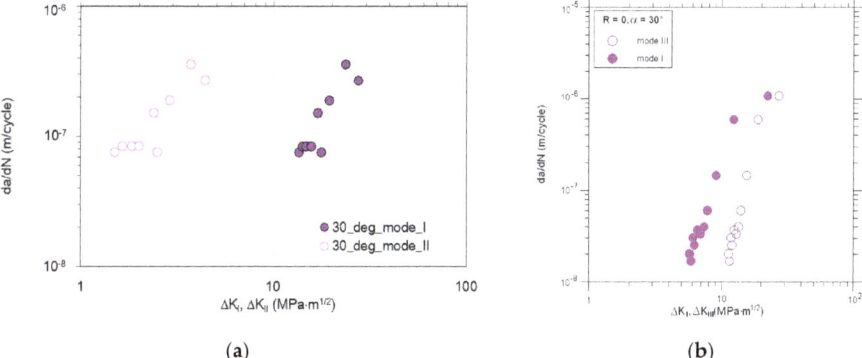

Figure 10. Comparison of the fatigue crack growth rates vs. ΔK vs. for $\alpha = \theta = 30°$: (a) Mode I and II; (b) mode I and III.

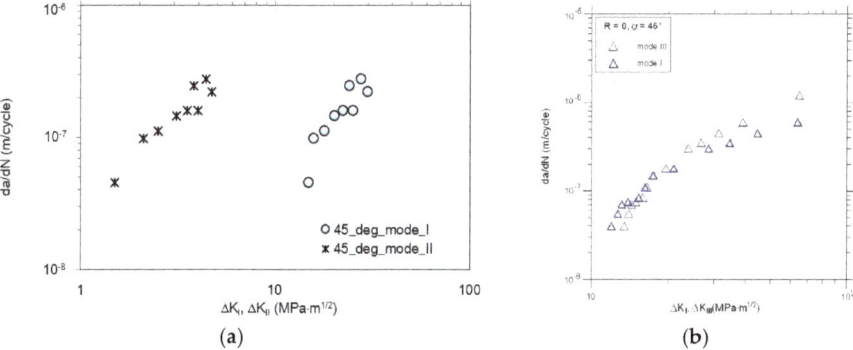

Figure 11. Comparison of the fatigue crack growth rates vs. ΔK for $\alpha = \theta = 45°$: (a) Mode I and II; (b) mode I and III.

It is worth to note that for increasing load angle (for all combinations of modes I + II and I + III) the higher fatigue lifetime of the specimen is observed. As it is also noticeable, for each case of mixed-mode I + II testing, the fatigue crack growth rates are higher for mode II in comparison with mode I. According to the KFFDs, the same slope of both curves seems to confirm a similar mechanism of fatigue crack growth for lower and higher mode mixity level. On the other hand, in the case of the combination mode I + III, it is noticeable that fatigue crack growth rate is higher for mode I in case of loading angle, $a = 30°$. However, increasing the mode mixity level and shear stress state for $a = 45°$ seems to change this tendency, for a relatively high level of $K > 18$ MPa·m$^{0.5}$. This phenomenon should be reflected in the topography of the fatigue-fracture surfaces.

3.1. Fatigue Crack Paths Study and Sem Analysis of Fracture Surfaces for Mixed-Mode I + III Loading

According to reference [26], the most widely used crack branching criterion is the Maximum Tangential Stress (MTS) criterion, where the initial angle of crack growth satisfies the following Equation (10):

$$K_I \sin\theta + K_{II}(3\cos\theta - 1) = 0 \tag{10}$$

and finally, allow determining the initial angle ψ of mixed-mode fatigue cracking:

$$\tan\left(\frac{\psi}{2}\right)_{1,2} = \frac{K_I}{4K_{II}} \pm \frac{1}{4}\sqrt{(K_I/K_{II})^2 + 8} \tag{11}$$

Figure 12 shows the crack propagation angles measured after experiments in cracked CTS specimens along with predictions based on MTS criterion.

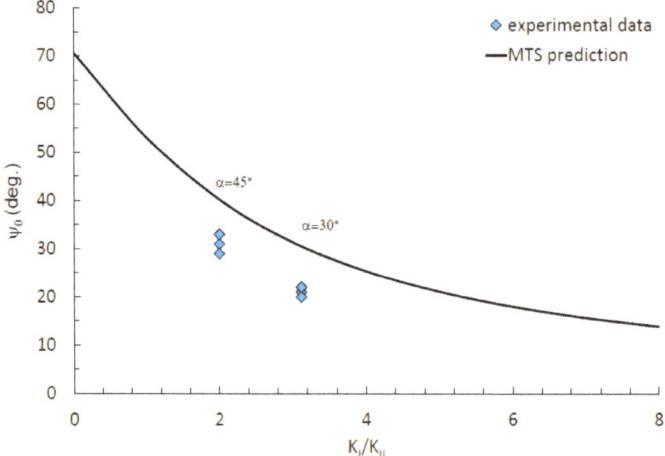

Figure 12. Crack initiation angles, according to MTS (Maximum Tangential Stress) criterion and experimental data for mode I + II.

As it is noticeable, the initial fatigue crack angles do not follow the MTS criterion, which suggests further modification of the crack initiation angles prediction models. This situation can be associated with material damage accumulation after 100 years of operation. In the case of long-term operated steels, this phenomenon should be further investigated.

On the other hand, fatigue tests at mixed I-III loading show that the smaller the amplitude of mode III loading (through deviation of the movable capture of specimens at the angles of 30 and 45 degrees), the higher the fatigue life of the specimens (6×10^4 and 5.5×10^4, respectively). Investigations of the topography of the fatigue-fracture surfaces of specimens show a significant effect of component III already at the macro level. During fatigue tests at mode I, fracture surfaces are usually formed normally oriented with respect to the direction of the acting stresses. Whereas, under the combined action of bending and torsion, the macro fracture surfaces of the specimens deviated from this orientation when approaching to their side surfaces. Moreover, the higher amplitude of the loading component K_{III}, the greater the angle of deviation of the fracture surface was observed (Figure 13a). Also, this higher amplitude caused the more enormous heights of the ridges in the main direction of crack growth formed on the fracture surfaces (Figure 13b–d). These ridges correspond to the interface of local parts of the fracture surfaces with different orientations relative to the normal orientation in the center of the section of the specimens along with their thickness. Such a stepwise transition from one local part of the fracture surface to another ultimately ensured its deviation at the macro level of the regarding observed near side surfaces of specimens.

The zones of the initial stage of the fatigue crack growth, its accelerated growth, when the angle of inclination of the fracture surfaces became more noticeable, and spontaneous growth zones were observed on the macro fracture surfaces of the specimens (Figure 13a). The fatigue crack growth features at the micro-level were analyzed at the center parts of the fracture surfaces of the specimens by their thickness and near their side surfaces.

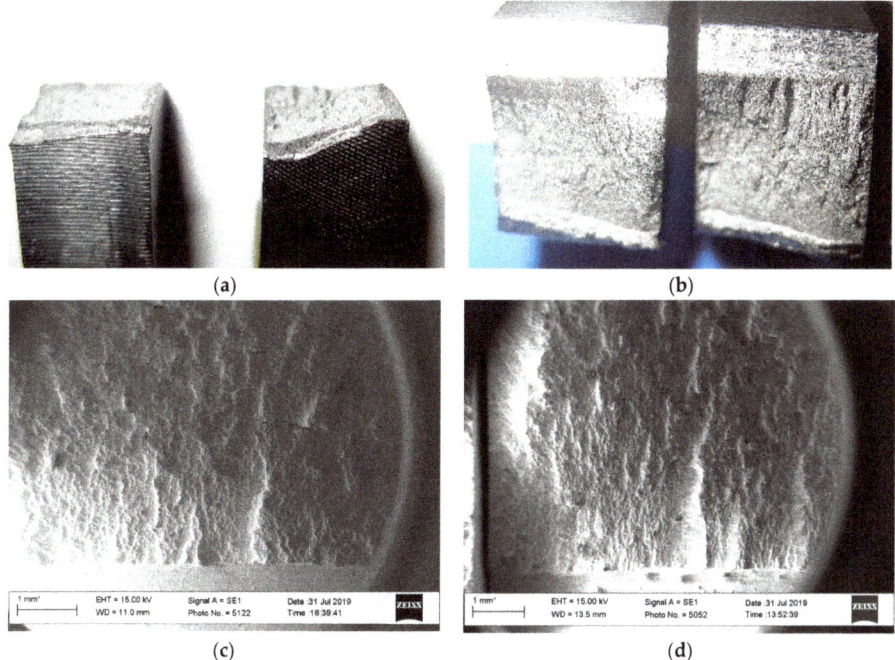

Figure 13. Profiles of the fracture surfaces of mode I-III test specimens: (**a**) Overview; (**b–d**) fracture features on the macro-level of the specimens.

A common feature of all fracture surfaces was their damaging due to the contact of the conjugated fracture surfaces in the loading cycle (Figure 14a), due to the crack closure effect. The fracture surface obtained at the minimum loading amplitude according to the mode III loading type (at 30 degrees) was the most damaging due to the crack closure effect. The remains of an undamaged fatigue relief were found on this fracture surface only along the centerline of the specimen thickness. This is due to the smallest height of the ridges at the interfaces of adjacent local parts on the fracture surfaces due to the gradual reorientation of their inclination to the fracture surface. It is clear that they were completely damaged by the friction of surfaces during the specimen test. Therefore, the residues of the typical fatigue relief (undistorted by the friction of surfaces during this specimen test) could be only observed in the deepenings at the base of the ridges on this fracture surface. In this case, the small fragments with typical relief of fatigue striations perpendicular to the main direction of crack growth were observed (Figure 14b).

Fracture surfaces of the specimens tested at higher loading amplitudes of K_{III} type (at 45 degrees) were more suitable for fractographic studies. The higher height of the ridges on the fracture surfaces connecting with their reorientation under the influence of the K_{III} type loading component enables for the preservation of a higher number of fracture areas, with typical fatigue relief observed against the background of artefacts formed during contact of the mating surfaces in each loading cycle. At both loading amplitudes under the K_{III} type, the typical fatigue features were observed on the fracture surfaces. In particular, among such features were the festoons elongated in the macro direction of crack growth. Within each of them, the same direction of local crack growth was observed, and fatigue striations oriented across them covered their surfaces (Figure 15). The influence of the load, according to K_{III} type, was manifested in the following. At the loading of specimens according to K_I type, the flat festoons delimiting the areas with locally unidirectional crack growth are usually observed [27]. However, under the influence of K_{III} type, loading the surfaces of each from the festoons revealed the fracture surfaces of specimens were curved. This was especially noticeable in the transitions between adjacent festoons. Obviously, during the testing of specimens according to the K_I–K_{III} type, the festoons were not merged by cleavage way (as it takes place after testing at the K_I type loading), but due to the plastic stretching of the bridges between adjacent festoons with the formation of curvilinear ridges between them. The role of the loading amplitude according to scheme K_{III} is almost levelled at the stage of accelerated crack growth with the expected increase of the spaces between fatigue striations.

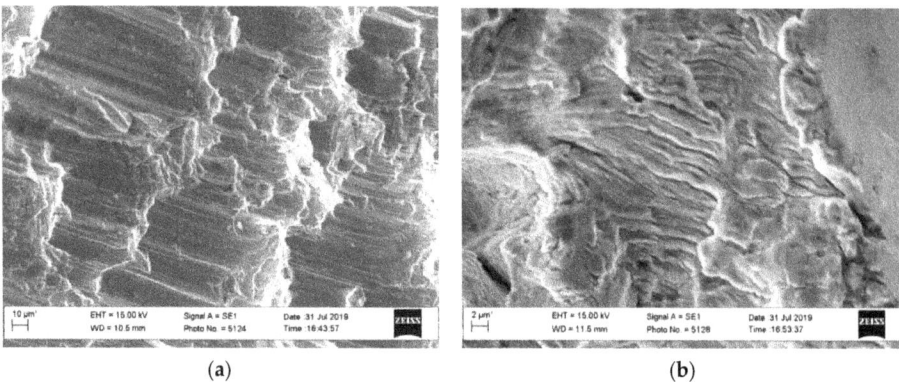

Figure 14. Typical illustrations of the fatigue-fracture surfaces of a specimen tested at a minimum amplitude of K_{III} type (30 degrees): (**a**) Traces of contact between the conjugated surfaces in the load cycle; (**b**) fatigue striations at the final stage within starting zone of the crack growth (at the crack length near 2 mm) in the middle of the specimen thickness.

The most interesting difference between the analyzed fracture surfaces from the observed specimens loading under K_I type consists of the following. With the specimens tested at the K_I–K_{III} loading type, the direction of crack propagation is changed with its growth into the deep of the cross-section of specimens. In the middle of their thickness, the traditional mutually perpendicular orientation of the fatigue striations regarding the main direction of crack growth is observed. Near the end of the starting fatigue zone, more and more areas were recorded on the fracture surfaces with other orientations (Figure 16a,b). The orientation of the fatigue striations gradually changed, and at approaching the side surfaces of the specimen, it could even become parallel to the main direction of crack growth. In other words, the crack propagation from the middle part of the specimen along their thickness in the directions to the side surfaces has begun.

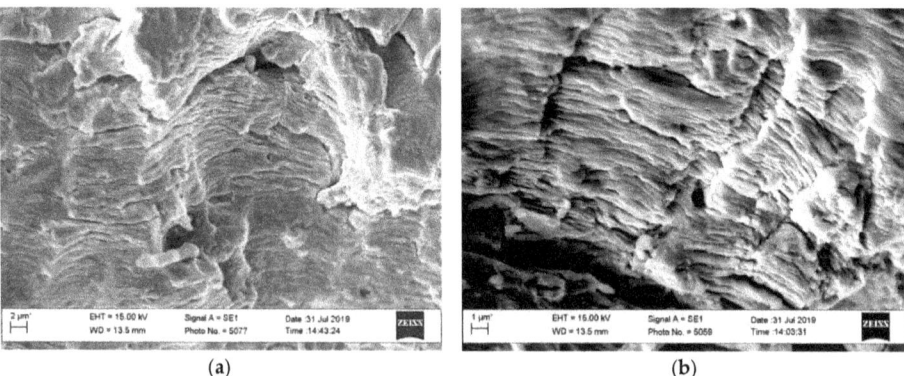

Figure 15. *Cont.*

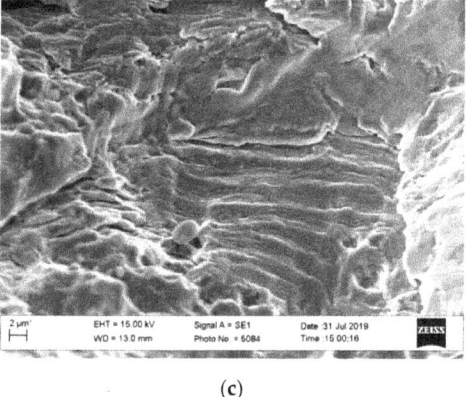

(c)

Figure 15. Fractograms in the center of the specimen thickness of the fracture surface: (**a**) Within the starting fracture zones; (**b**–**c**) within the zones of accelerated crack growth of the specimens, K_I–K_{III} with load angle $a = 45°$.

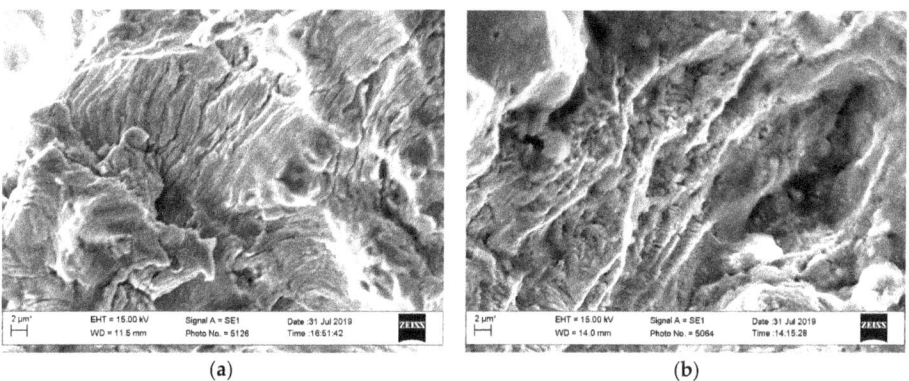

(a) (b)

Figure 16. Fractograms of specimens obtained at a distance between 1 and 3 mm from their side-surfaces at the final stage of the starting fracture zones of specimens loaded at K_I–K_{III} type: (**a**) Up to 30 degrees by K_{III}; (**b**) up to 45 degrees by K_{III}.

The characteristic elements of a ductile fracture with a typical dimple relief were observed within the zone of the spontaneous crack growth of tested specimens (Figure 17). It was only noted the significant contribution of shear deformations to the formation of dimples since parabolic ridges delineated them.

It was concluded that regardless of the amplitude of the K_{III} type loading, the initiation of the fatigue crack growth in the specimens begins from stress concentrators in the middle of their thickness due to K_I type stress effect. Up to a certain depth of crack growth, their propagation rates along this centerline were significantly higher compared to realising ones with approaching to the side surfaces of the specimens. Nevertheless, at the final stage of the starting zones of fatigue crack growth, the local directions of the crack growth on the fracture surfaces deviate away from the centerline (along with the thickness of the specimens) due to K_{III} type stress effect. In this case, the reorienting of the crack growth direction in the direction of the side surfaces of the specimens was taking place. It means the role of the K_{III} type loading is to the reorientation of the fatigue crack growth direction.

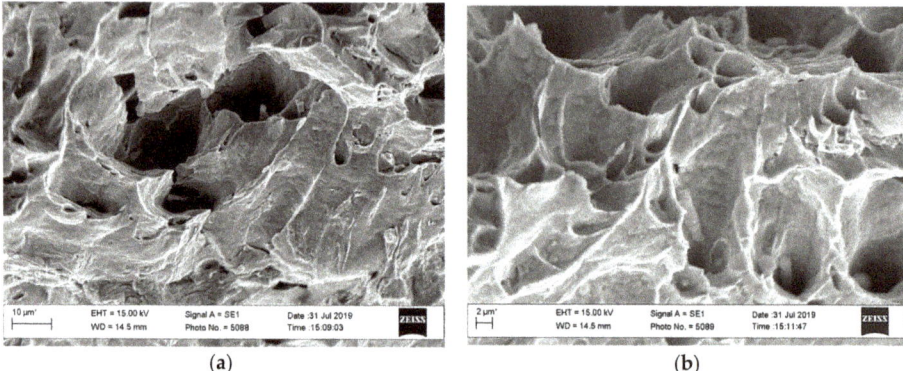

Figure 17. Fractograms of the specimen loaded according to the K_I–K_{III} scheme (with the K_{III} type loading up to 45 degrees), at the stage of spontaneous crack growth: (**a**) 10 µm; (**b**) 2 µm.

4. Conclusions

The following conclusions from the test results for fatigue crack growth in long operated bridge steel under mixed-mode (I + II, I + III) loading conditions can be drawn:

- Under mixed modes condition, I + II the fatigue lifetime increase with the increase of loading angle θ value. It is caused by the decreasing of the ΔK_I.
- Increase of the angle α determining a ratio of the torsional moment to the bending moment causes a decrease of the fatigue life.
- For $\alpha = 45°$ and mode I-III, a higher crack growth rate is observed for mode I, which goes into mode III domination. Fatigue crack closure effect evidence is noticeable under mixed-mode I + III configuration, the sliding mode fracture type is observed. However, it is more likely that is caused by damaged fracture surfaces during multiaxial loading.
- The initiation angle is lower than predicted by MTS criteria. It might be caused by the material degradation—microstructure and additional internal stresses, strains (the similar effect was observed by authors in physically prestrained specimens from modern P355NL1 steel [27] and in reported results published in [28]) after 100 years operating time, but this phenomenon should be further investigated.

Author Contributions: Conceptualization, G.L and M.S.; methodology, J.C. and A.D.J.; validation, G.L., M.S. and R.M.; formal analysis, G.L., H.K., O.S., D.R.; investigation, G.L., D.R. and M.S.; resources, G.L.; data curation, G.L. and M.S.; writing—original draft preparation, J.C., H.K., O.S.; writing—review and editing, M.S. and R.M.; visualization, A.D.J.; supervision, G.L., A.D.J., J.C.; project administration, G.L.; funding acquisition, G.L. All authors have read and agreed to the published version of the manuscript.

Funding: This work was supported in a part by grant number 2018/02/X/ST8/02041 (02NA/0001/19) financed by the Polish National Science Centre (Narodowe Centrum Nauki, NCN). Additionally, this research was supported by the research unit CONSTRUCT (POCI-01-0145-FEDER-007457, UID/ECI/04708/2019, FEDER, and FCT/MCTES) and to the SciTech-Science and Technology for Competitive and Sustainable Industries, R & D project NORTE-01-0145-FEDER-000022 financed by Programa Operacional Regional do Norte ("NORTE2020"), through Fundo Europeu de Desenvolvimento Regional (FEDER).

Conflicts of Interest: The authors declare no conflict of interest.

References

1. Kuhn, B.; Lukić, M.; Nussbaumer, A.; Guenther, H.P.; Helmerich, R.; Herion, S.H.K.M.; Bucak, Ö. *Assessment of Existing Steel Structures: Recommendations for Estimation of Remaining Fatigue Life*; Office for Official Publications of the European Communities: Brussels, Belgium, 2008.

2. Moreno, J.; Valiente, A. Assessment of the reference stress method for J-integral estimation of cracked riveted beams of an old wrought iron. *Eng. Fail. Anal.* **2008**, *15*, 194–207. [CrossRef]
3. Mayorga, L.G.; Sire, S.; Calloch, S.; Yang, S.; Dieleman, L.; Martin, J.L. Fast characterization of fatigue properties of an anisotropic metallic material: Application to a puddled iron from a nineteenth century French railway bridge. *Procedia Eng.* **2013**, *66*, 689–696. [CrossRef]
4. Walker, R. The production, microstructure, and properties of wrought iron. *J. Chem. Educ.* **2002**, *79*, 443. [CrossRef]
5. Gordon, R.; Knopf, R. Evaluation of wrought iron for continued service in historic bridges. *J. Mater. Civ. Eng.* **2005**, *17*, 393–399. [CrossRef]
6. Mayorga, L.G.; Sire, S.; Plu, B. Understanding fatigue mechanisms in ancient metallic railway bridges: A microscopic study of puddled iron. *Procedia Eng.* **2015**, *114*, 422–429. [CrossRef]
7. Lesiuk, G.; Kucharski, P.; Correia, J.A.; De Jesus, A.M.P.; Rebelo, C.; da Silva, L.S. Mixed mode (I + II) fatigue crack growth in puddle iron. *Eng. Fract. Mech.* **2017**, *185*, 175–192. [CrossRef]
8. Lesiuk, G.; Kucharski, P.; Correia, J.A.F.O.; De Jesus, A.M.P.; Rebelo, C.; Da Silva, L.S. Mixed mode (I + II) fatigue crack growth of long term operating bridge steel. *Procedia Eng.* **2016**, *160*, 262–269. [CrossRef]
9. Lesiuk, G.; Correia, J.; Smolnicki, M.; De Jesus, A.; Duda, M.; Montenegro, P.; Calcada, R. Fatigue Crack Growth Rate of the Long Term Operated Puddle Iron from the Eiffel Bridge. *Metals* **2019**, *9*, 53. [CrossRef]
10. Li, Z.X.; Chan, T.H.; Ko, J.M. Fatigue analysis and life prediction of bridges with structural health monitoring data—Part I: Methodology and strategy. *Int. J. Fatigue* **2001**, *23*, 45–53. [CrossRef]
11. Moreno, J.; Valiente, A. Cracking induced failure of old riveted steel beams. *Eng. Fail. Anal.* **2006**, *13*, 247–259. [CrossRef]
12. Ooi, E.T.; Yang, Z.J. A hybrid finite element-scaled boundary finite element method for crack propagation modelling. *Comput. Methods Appl. Mech. Eng.* **2010**, *199*, 1178–1192. [CrossRef]
13. Panasyuk, V.V.; Schuller, M.; Nykyforchyn, H.M.; Kutnyi, A.I. Corrosion-hydrogen degradation of the shukhov lattice construction steels. *Procedia Mater. Sci.* **2014**, *3*, 282–287. [CrossRef]
14. Lesiuk, G.; Szata, M.; Bocian, M. The mechanical properties and the microstructural degradation effect in an old low carbon steels after 100-years operating time. *Arch. Civ. Mech. Eng.* **2015**, *15*, 786–797. [CrossRef]
15. Nykyforchyn, H.; Lunarska, E.; Tsyrulnyk, O.T.; Nikiforov, K.; Genarro, M.E.; Gabetta, G. Environmentally assisted "in-bulk" steel degradation of long term service gas trunkline. *Eng. Fail. Anal.* **2010**, *17*, 624–632. [CrossRef]
16. Haghani, R.; Al-Emrani, M.; Heshmati, M. Fatigue-prone details in steel bridges. *Buildings* **2012**, *2*, 456–476. [CrossRef]
17. Deng, L.; Wang, W.; Yu, Y. State-of-the-art review on the causes and mechanisms of bridge collapse. *J. Perform. Constr. Facil.* **2015**, *30*, 04015005. [CrossRef]
18. Lesiuk, G.; Katkowski, M.; Duda, M.; Królicka, A.; Correia, J.A.F.O.; De Jesus, A.M.P.; Rabiega, J. Improvement of the fatigue crack growth resistance in long term operated steel strengthened with CFRP patches. *Procedia Struct. Integr.* **2017**, *5*, 912–919. [CrossRef]
19. Richard, H.A. A new compact shear specimen. *Int. J. Fract.* **1981**, *17*, R105–R107.
20. Lewandowski, J.; Rozumek, D. Cracks growth in S355 steel under cyclic bending with fillet welded joint. *Theor. Appl. Fract. Mech.* **2016**, *86*, 342–350. [CrossRef]
21. Rozumek, D.; Marciniak, Z.; Lesiuk, G.; Correia, J.A.; de Jesus, A.M. Experimental and numerical investigation of mixed mode I + II and I + III fatigue crack growth in S355J0 steel. *Int. J. Fatigue* **2018**, *113*, 160–170. [CrossRef]
22. Rozumek, D.; Marciniak, Z.; Lachowicz, C.T. The energy approach in the calculation of fatigue lives under non-proportional bending with torsion. *Int. J. Fatigue* **2010**, *32*, 1343–1350. [CrossRef]
23. Sih, G.C. Strain-energy-density factor applied to mixed mode crack problems. *Int. J. Fract.* **1974**, *10*, 305–321. [CrossRef]
24. Harris, D.O. Stress intensity factors for hollow circumferentially notched round bars. *J. Fluids Eng.* **1967**, 49–54. [CrossRef]
25. Chell, G.G.; Girvan, E. An experimental technique for fast fracture testing in mixed mode. *Int. J. Fract.* **1978**, *14*, R81–R83.
26. Erdogan, F.; Sih, G.C. On the crack extension in plates under plane loading and transverse shear. *J. Basic Eng.* **1963**, *85*, 519–525. [CrossRef]

27. Ferreira, J.; AFO Correia, J.; Lesiuk, G.; Blasón González, S.; Gonzalez, R.; Cristina, M.; Fernández-Canteli, A. Pre-Strain Effects on Mixed-Mode Fatigue Crack Propagation Behaviour of the P355NL1 Pressure Vessels Steel. In Proceedings of the ASME 2018 Pressure Vessels and Piping Conference, Prague, Czech Republic, 15–20 July 2018.
28. Student, O.Z.; Dudziński, W.; Nykyforchyn, H.M.; Kamińska, A. Effect of high-temperature degradation of heat-resistant steel on the mechanical and fractographic characteristics of fatigue crack growth. *Mater. Sci.* **1999**, *35*, 499–508. [CrossRef]

© 2020 by the authors. Licensee MDPI, Basel, Switzerland. This article is an open access article distributed under the terms and conditions of the Creative Commons Attribution (CC BY) license (http://creativecommons.org/licenses/by/4.0/).

MDPI
St. Alban-Anlage 66
4052 Basel
Switzerland
Tel. +41 61 683 77 34
Fax +41 61 302 89 18
www.mdpi.com

Materials Editorial Office
E-mail: materials@mdpi.com
www.mdpi.com/journal/materials

www.ingramcontent.com/pod-product-compliance
Lightning Source LLC
LaVergne TN
LVHW071953080526
838202LV00064B/6736